Tropical Deforestation and Species Extinction

IUCN – THE WORLD CONSERVATION UNION

There are thousands of environmental organizations around the world – but the World Conservation Union is unique. It is the only one whose membership includes both governments and non-governmental organizations, providing a means to work together to achieve effective conservation action on the ground. IUCN's mission – to provide leadership and promote a common approach for the World Conservation Movement – translates into a practical aim to establish links between development and the environment that will provide a lasting improvement in the quality of life for people all over the world.

The Union's work is increasingly decentralized and is carried forward by an expanding network of regional and country offices, located principally in developing countries and working in close partnership with headquarters. Specialist programmes, covering themes such as ecology, species conservation, habitat and natural resource management, and the relationship between population and the environment, draw on scientific input from around the world. They develop strategies and services to address today's critical international environment and development issues – and seek to anticipate and address those of tomorrow. This work is further strengthened via a network of six Commissions, drawing together dedicated and highly qualified volunteers into project teams and action groups – for results.

THE IUCN COMMISSION ON ECOLOGY

The IUCN Commission on Ecology (COE) serves as the Union's source of technical advice for translating knowledge of ecological processes into practical action for conservation, sustainable management and restoration, in particular areas degraded by human action. The IUCN programmes on Forest Conservation, Marine Conservation, Global Change and Wetlands are guided by the Commission on Ecology, whose members serve on advisory committees and in working groups associated with these programmes.

THE IUCN FOREST CONSERVATION PROGRAMME

The IUCN Forest Conservation Programme (formerly the IUCN Tropical Forest Programme) coordinates and reinforces activities of the IUCN members and Secretariat which deal with forests. The Programme focuses on the conservation of species and ecological processes, and on investigating and promoting sustainable use of the resources of these forests.

The Programme includes international and national policy initiative strategies as well as field projects addressing selected problems in managing the world's most biologically significant forests. These selected projects put the World Conservation Strategy into action by reconciling the requirements of conservation with national development and the needs of the people living in forest areas. Special emphasis is given to the development of compatible uses for buffer zones around national parks and reserves.

IUCN develops its positions and policies on the basis of the concerns and information communicated by members. Trends identified by monitoring activities, and the feedback from numerous field projects. Data on species on plants and animals, and on forest sites which are important for biological and ecosystem conservation, are held by the World Conservation Monitoring Centre in Cambridge, UK.

This series of publications from the Forest Conservation Programme, in conjunction with regular meetings, enables IUCN to communicate policies and technical guidance to governments, major international institutions, development planners, and conservation professionals. The Programme works closely with development assistance agencies, governments and NGOs, to ensure that conservation priorities are adequately addressed in their activities.

The Forest Conservation Programme receives generous financial support from the Government of Sweden.

Tropical Deforestation and Species Extinction

Edited by
T.C. WHITMORE
and
J.A. SAYER

CHAPMAN & HALL
London · New York · Tokyo · Melbourne · Madras

Published by Chapman & Hall, 2–6 Boundary Row, London SE1 8HN

Chapman & Hall, 2–6 Boundary Row, London SE1 8HN, UK

Chapman & Hall, 29 West 35th Street, New York NY10001, USA

Chapman & Hall Japan, Thomson Publishing Japan, Hirakawacho Nemoto Building, 6F, 1–7–11 Hirakawa-cho, Chiyoda-ku, Tokyo 102, Japan

Chapman & Hall Australia, Thomas Nelson Australia, 102 Dodds Street, South Melbourne, Victoria 3205, Australia

Chapman & Hall India, R. Seshadri, 32 Second Main Road, CIT East, Madras 600 035, India

First edition 1992

© 1992 International Union for Conservation of Nature and Natural Resources

Disc conversion in 10/12 pt Sabon by Columns, Reading, Berkshire
Printed in Great Britain by St Edmundsbury Press, Bury St Edmunds, Suffolk

ISBN   0 412 45520 X

Citation: Whitmore, T.C. and Sayer, J.A. (1992) *Tropical Deforestation and Species Extinction*. Chapman & Hall, London. xvii + 147 pp.

Also available from IUCN Publications Services Unit, 219c Huntingdon Road, Cambridge, CB3 0DL, UK.

The designations of geographical entities in this book, and the presentation of the material do not imply the expression of any opinion whatsoever on the part of IUCN or other participating organizations concerning the legal status of any country, territory or area, or of its authorities, or concerning the delimitation of its frontiers or boundaries.

A catalogue record for this book is available from the British Library

Library of Congress Cataloging-in-Publication data available

Contents

Contributors

G.G. BROWN Department of Agronomy, University of Wisconsin, Wisconsin, USA. (present address: CP6588 Barão Geraldo, Campinas, São Paulo – 13032, Brazil)

K.S. BROWN Instituto de Biologia, Universidade Estadual de Campinas, São Paulo, Brazil

V.H. HEYWOOD IUCN – Kew, Richmond, Surrey, UK

A.D. JOHNS Wildlife Conservation International, New York Zoological Society, Bronx Zoo, New York, USA. (present address: Makerere University Biological Field Station, PO Box 10066, Kampala, Uganda)

R.J. JOHNS Royal Botanic Gardens, Kew, Richmond, Surrey, UK

W.V. REID World Resources Institute, Washington DC, USA

J.A. SAYER Forest Conservation Programme, International Union for Conservation of Nature and Natural Resources, Gland, Switzerland

D. SIMBERLOFF Department of Biological Sciences, Florida State University, Florida, USA

S.N. STUART Species Survival Commission, International Union for Conservation of Nature and Natural Resources, Gland, Switzerland

T.C. WHITMORE Geography Department, University of Cambridge, Cambridge, UK

Preface

IUCNs Forest Conservation Programme works at two levels to conserve the biological diversity of the world's forests. First, it has on-the-ground projects in those forests, mainly in the tropics, where the fauna and flora is at greatest risk; second, it influences decisions at the national and international political level, where the fate of forests is very often determined. IUCN encourages richer nations to support forest conservation in poorer nations and considerable international funding is now available for this purpose. But the resources available are still inadequate to meet the challenge. It is therefore of great importance that the funding that is available is deployed in the most effective possible way. This can only be done on the basis of a thorough understanding of the processes which are responsible for the loss of forests and their associated species.

Estimates of rates of tropical deforestation vary enormously and the implications of this loss for species extinction are only beginning to be understood. Many people have asked IUCN to comment on the numerous conflicting estimates of species extinction and some would like us to come up with a firm and definitive figure for the number of species which are being lost in a given period of time. The data available would not enable this to be done with any reasonable degree of scientific credibility and we have not attempted to do so in this book. Our aim has rather been to request leading researchers in the field to present their work in such a way that we will be able to better understand the processes that are at work. We believe that this understanding of the processes is more important than quantifying the consequences in enabling us to make better decisions on how conservation resources can be most effectively deployed.

The chapters in this book were presented at a workshop held during the 18th General Assembly on IUCN held in Perth, Australia in November and December 1990. Over 100 specialists in tropical forest conservation and management attended the workshop and their views are reflected in the revised versions of the papers that are presented here. One of the products of the workshop was a resolution adopted by consensus by the 1500 representatives of IUCNs 650 member organizations who were present in Perth. The text of this resolution was as follows.

18.20 Tropical Forest Alteration and Species Extinctions

RECOGNIZING that recent estimates by the Food and Agriculture Organization of the United Nations show that rates of deforestation in the tropics are at least 50 per cent higher than had previously been thought and that there is also an increase in areas of primary forest being utilized for timber;

RECOGNIZING that all forest alteration leads to local extinction and genetic erosion of species;

NOTING that tropical forests are generally acknowledged to be the most species rich environment on earth;

NOTING WITH APPRECIATION the conclusions and recommendations of the workshop on Tropical Deforestation Rates and their Implications for Species Extinction, held at the present General Assembly;

CONCERNED that the rate of acquisition of new national parks and equivalent totally protected areas has declined, that less than 5 per cent of tropical moist forests are included in such areas and that this is inadequate to conserve the full range of biological diversity of tropical forests;

FURTHER CONCERNED that past estimates of species loss from tropical deforestation have been based upon simple species-area relationships and that these may underestimate the problem by failing to take account of patterns of forest loss and particularly the risks associated with forest fragmentation;

AWARE that by optimally locating protected areas in regions of high species diversity and endemism the value of these areas for species conservation can be greatly enhanced;

The General Assembly of IUCN – The World Conservation Union, at its 18th Session in Perth, Australia, 28 November–5 December 1990:

1. CALLS UPON all nations concerned with the impact of tropical forest loss or change on species richness and diversity to:
   a urgently undertake surveys to identify additional sites of critical importance for conservation of biological diversity due to the endemism, representivity, richness or scarcity of their species, and wherever possible accord total protection to these sites;
   b relocate harvesting from primary forests to secondary forests and to tree plantations in previously deforested areas or, where this is not possible, work towards the development and introduction of forest harvesting systems which favour natural species diversity;
   c manage the national forest estate so as to optimize the conservation of natural species diversity by appropriate use of managed forests to buffer totally protected areas and provide corridors to link them;

2. RECOMMENDS nations whose tropical forests are already severely depleted, or whose tropical forest areas were never large, to protect and maintain all forest remnants in the interests of conserving their heritage of wild species;

3. URGES the industrialized nations to support the efforts of nations to conserve their tropical forests, through appropriate development aid

and measures to decrease the foreign debts and improve the trade relations of these countries;

4. REQUESTS the Director General of IUCN to support, within the resources available, as well as through contacts with members and potential sources of funding, those nations and organizations striving to achieve development objectives in harmony with the maintenance of the biological diversity of tropical forests.

The extinction of species that occurs as a result of tropical deforestation is one of the most serious problems confronting the conservation community. We hope that this book will increase understanding of the complexity of the issues involved and will lead to greater and more effective international action to stem the loss.

Tim Whitmore, Cambridge, UK.
Jeff Sayer, Gland, Switzerland.

Foreword

MARTIN W. HOLDGATE
Director General, IUCN – The World Conservation Union

The direct pressures of mounting human numbers, and the less direct but equally profound changes in the biosphere that result from human perturbation of biogeochemical cycles, are altering the course and character of biological evolution. The world community is committed to development that meets the needs of present and future human populations. That development will only succeed if it is based on the conservation of the essential natural systems and resources of the planet, and uses them in a sustainable way.

Forest is the potential vegetation of much of the land surface of the Earth. Forests maintain soils and nutrient and water cycles, and are major reservoirs of biological diversity. They are the habitats of at least half, and probably more, of all species. They are also the homes of many peoples who have developed life styles that are in harmony with the forest habitat.

But, inevitably, forests are also at risk from development. The growth in human numbers is a direct consequence of advances in agriculture, much of it on land cleared from forest. Urban and industrial civilizations are, for the most part, inimical to the forest environment. As human numbers mount from today's 5.4 billion towards the projected 10 – 12 billion of the latest UN estimates (even if other factors intervene to prevent these possibly unsupportable totals being attained), forests will come under further pressure. This will be most acute in the tropics, where population growth will be concentrated and where the largest areas of diverse forest remain.

The world's conservation community has a duty to help steer humanity towards a sustainable future. This cannot be achieved by seeking to obstruct all development that takes cultivable land to grow food for the billions that will inevitably be added to the world population, or by resisting all conversion to provide sites for new industries and urban centres. It will require a conscious process of survey, inventory, and evaluation that tries to ensure that every hectare of land is used in an optimal way – a way that meets needs today and in the future.

This book has been produced as a contribution to that process. It arose from the debate over rates of deforestation and species extinction in the tropics – and many requests to IUCN for an evaluation of these rates and their significance. But, as the book came to be prepared it also became evident that these issues needed a wider context. For this reason, while the chapters in this volume discuss rates of forest loss and species extinction, they also consider how forests should be conserved and managed for human benefit.

Forests provide, and will continue to provide, resources of immense value for the human future. Their role as a part of the human life-support system, regulating local climates, water flow, and nutrient cycles, must be recognized, valued, and safeguarded. So must their role as reservoirs of biological diversity and the habitat of species which, as the World Charter for Nature (Burhenne and Irwin, 1983) emphasized, deserve respect regardless of their usefulness to humanity. And their value for the products they provide for human use is another crucial attribute. Timber, nuts and other fruits, meat, fibre, latex and medicines are all produced from forests and if valued correctly and added to the less-tangible values of life support, may commonly aggregate to a greater economic worth than the same land can provide when converted into farmlands or urbanization.

One task of conservation is thus to speak for the value of forests, as habitats of species, and as providers of services. To do this, we need scientific understanding of forests as diverse systems. Unless we understand the factors that govern their structure and dynamics, we are unlikely to manage them correctly: and management regimes have to be designed to fit the circumstances of particular ecosystems. The human vision of forests is commonly one of uniformity: to the European settler in the tropics all forests were 'bush' to be cleared or logged, and treated much the same way as the temperate woodlands from which those people came. The management systems imported by northern hemisphere foresters have not proved successful in very different kinds of setting, at least without substantial adaptation. As the chapter by A.D. Johns, in this book, makes clear, we have a long way to go before we arrive at management regimes for tropical forests which are truly sustainable, maintaining their capacity to meet a broad spectrum of human uses now and for an indefinite number of future generations.

We are even more uncertain about how to design forest conservation and management so as to maintain biological diversity. Indeed, there are major uncertainties over the very nature of that diversity: over how many species, of what major taxonomic groups, live in what types of forest in which countries. Many of the figures we use are best estimates, derived through an explicit process of extrapolation, but often quoted without the statistical margins that should be attached to such estimates. Similarly, it is a matter of common sense that the destruction of some 18 million hectares of closed tropical forest annually must be accompanied by the loss of some of the species dependent on those forests — but the precise statement of how many species are being lost now, or put at risk in the future is another matter of estimation and extrapolation.

Two kinds of method have been, and will continue to be, used to estimate species loss. One is direct observation. The chapters in this book by Heywood and Stuart and by Whitmore and Sayer point out that we have partially reliable data on species loss for birds and mammals, but for few other taxonomic groups. Between 1600 and the present day, we know we have lost 83 species of mammal and 113 species of bird, plus 384 kinds of vascular plant. As these authors, and also Reid, point out, this is a substantial acceleration of average natural rates, but it is far less than the inferred rates of extinction from calculations based on assumptions about species association with forest area, and the rates of loss of forest. The 'modelling' approach is illustrated in the chapters by Simberloff and by Reid, and should be taken as a warning in two senses. First, it may well indicate that observed extinctions of the more conspicuous and better studied groups are a poor guide to the seriousness of the problems confronting us; and second, the gap between prediction and observation is a testimony to the need for much more scientific effort in this field.

One area for such study is well illustrated in the chapter by Brown and Brown. The coastal forests of Brazil have been reduced in area as severely as any tropical forest type in the world. According to calculation, this should have led to considerable species loss. Yet no known species of its old, largely endemic, fauna can be regarded as extinct. Genetic erosion has undoubtedly taken place, and the reduced, remnant populations may be much more vulnerable to future change, but the study illustrates the need for very careful field documentation to compare with calculation in this and other situations.

There are other major problems we need to recall. Despite the impression of permanence and robustness they may convey to the casual visitor, the world's great forests are not static. Ten thousand years ago, as the Earth emerged from the last glacial epoch, the forests of South America and Africa were fragmented. The great, continuous expanses, now to be seen in Zaire and the Amazon basin are relatively recent phenomena. Climate change, now projected to combine rate and magnitude in an unprecedented way, may well change the limits of tolerance of species by several hundreds of kilometres in less than a century. Rainfall patterns may alter. Sea level is likely to rise by around 20 centimetres in 40 years, and may increase by over half a metre before the year 2100. The ecological impacts are likely to be profound, superimposed as they will be on intense direct human pressures.

As the chapters by Whitmore and Sayer, by Reid and by Simberloff all say in different ways, there are grave doubts about how far islands of forest within protected areas can serve as reservoirs for the diversity of forest species. It will be even more difficult to maintain this diversity if climate change makes some of those protected areas no longer suitable to support the species that at present live in them – and yet, because the adjacent lands have all been converted for intensive human use, there are no corridors or adjacent habitats into which species can spread, or through which they can disperse.

IUCN, as the World Conservation Union, is concerned with the whole spectrum of issues that these chapters raise. The Union seeks to ensure that the world's biological diversity is safeguarded. Protected areas play, and will continue to play, a key role, but given population pressures, climate change, and the difficulties in managing these areas effectively, we cannot rely on them alone or treat them in isolation. IUCN is committed to work for species survival, and this has to be done by integrating nature conservation with land management for human use, outside as well as within protected areas. The Union recognizes the need for development and resource use that meets urgent human needs, and this means the development of methods for sustainable forest use, and ensuring that only those areas that are more productive and beneficial under other land uses are converted.

All these tasks demand a foundation of science. The duty of IUCN is to work to improve understanding in the various fields covered in this book. We need to increase the effort devoted to studies of the relationship between habitat and species diversity, to securing better measurements of habitat change and species loss, to more observations that will validate and improve the models on which we at present depend, and to formulating better management methods. We need to make the case for more commitment to the fundamental, and now neglected, field of biological taxonomy, without which we shall not be able to characterize the species we are seeking to conserve.

The chapters in this book should be seen as an interim statement of the position. The fact that they do not wholly agree with one another, or with other statements in the literature, should be seen as evidence of the need for more work, not as refutation of discordant statements. The situation is uncertain in many respects. Despite the revised figures quoted by Whitmore and Sayer, we still do not know precisely how much tropical forest we are losing each year. What we do know is that there are compelling grounds for strengthening our efforts for tropical forest conservation and sound management. A policy of doing nothing – or doing no more than at present – will certainly lead to a quite avoidable loss of biological diversity, and will impose an equally avoidable cost on humanity, now and in the future.

REFERENCE

Burhenne, W.E. and Irwin, W.A. (1983) The World Charter for Nature: a background paper. Buiträgs zud umwelt-gestaltung A90. Erich Schmid Verlag, Berlin.

Deforestation and species extinction in tropical moist forests

T.C. WHITMORE and J.A. SAYER

INTRODUCTION

Mankind has been destroying forests for millennia, ever since the discovery of agriculture (Williams, 1989). The assault on the temperate forests of North and South America and Australasia during the seventeenth to nineteenth centuries, which followed on European expansion, has been followed today by an assault on tropical rain forests.

The deforestation of tropical lands can have both local and global consequences. Locally, climates may become more extreme, soils may suffer physical and chemical deterioration, and hydrological balances may be perturbed. Massive deforestation, altering albedo and regional atmospheric water balance, could affect weather patterns, and there is particular concern at the possible contribution to atmospheric warming of the addition of carbon dioxide into the atmosphere from the burning or decomposition of the biomass. Perhaps the single greatest cause for concern over the loss of tropical forests is that there is a considerable body of evidence to suggest that it is leading to unprecedented loss of the biological diversity that these forests contain (Myers, 1983; Raven, 1987). Our knowledge of the numbers, distribution, status and ecology of tropical forest species is so poor that the true dimensions of this problem are only beginning to emerge. In the absence of the knowledge that would enable us to manage forests in a deliberate way to conserve their species diversity it must be prudent to retain as much forest as possible under a region of minimal human disturbance.

This protectionist approach does not necessarily apply when one is addressing the issue of conserving other forest values. Many of both the local and the global environmental services of tropical forests can be provided by modified vegetation cover. Timber plantations and tree crops can store substantial quantities of carbon. They can also, as can plantation crops and carefully managed agricultural systems, moderate climate and regulate water supply in a manner similar to natural forests. Even the lengthy fallow phase of traditional forms of shifting agriculture carries few long-term environmental risks.

The rate of tropical forest loss is so rapid, and the concentration of the world's species in these ecosystems so great (Wolf, 1987) that the view has been expressed (e.g. Ehrlich and Ehrlich, 1981; Simberloff, 1986) that a significant proportion of all species of plants and animals are likely to become extinct in the next few decades, perhaps as many as 25–30% by AD2000 (Myers, 1983). Other authors (Simon, 1986; Lugo, 1988) have contended that there is little documented evidence of species extinction, that many species survive deforestation, and that the risks inherent in tropical deforestation have been exaggerated.

Whatever the rate of extinction may be, there is no doubt that any disturbance of natural forests, whether man-made or natural, will alter the relative frequency of animal and plant species and severe disturbance is likely to cause extinctions.

In response to world-wide concern at the prospect of large numbers of species extinctions, the funds available for the conservation of biological diversity in the tropics are increasing rapidly. It is clearly important that these resources be optimally deployed. This requires a clear understanding of the whole issue, which in its turn is dependent on good data on the extent of remaining forest, the rates and locations of clearance, and the actual relationship between forest clearance and species loss. These matters are all currently subjects of intense debate. Interested parties include biologists, FAO, IUCN, WWF and aid agencies.

RATES OF TROPICAL MOIST FOREST DESTRUCTION

Various problems are encountered in any attempt to quantify forest cover and its change over time. First, it is necessary to define what is meant by forest, and which particular types of forest are to be considered. Second, forests change either due to complete destruction (that reduces their area), or because of various forms of disturbance (that change biomass, structure and species composition, and are hard to quantify). It is necessary to specify whether destruction, disturbance or both have occurred. Third, until the recent advent of satellite remote sensing technology it was impossible to achieve an extensive objective record of forest cover at a single moment in time.

Differences between published figures on tropical deforestation arise largely because these problems have been ignored or tackled in different ways. The first, and still the most complete, survey which addressed these issues and dealt with all countries consistently was published in 1981 by FAO to a 1980 dateline (FAO, 1981). This was revised by FAO (1988) to correct errors and extend the cover to some small countries previously omitted, but the 1980 dateline was retained. The later report significantly raised deforestation estimates for some important countries. Lanly (1992) has provided a commentary. The FAO survey is the only one which gives global and individual country figures for both forest disturbance and forest clearance.

A distinction was made by FAO between closed forests and open forests (or woodlands). The latter are '. . . forest/grassland formations with a continuous grass layer in which the tree synusia covers more than 10 percent of the ground', crown cover percentage varies and 'in some woodlands the trees may cover the ground completely, as in closed forests'. The distinction is 'more of an ecological than physiognomic type' (FAO, 1988, p.15). Closed and open forests can be separated on aerial photographs and satellite images.

Since the last century it has been common to recognize three major kinds of closed tropical forests: rain forests, monsoon (or seasonal) forests and thorn forests (Schimper, 1903), the occurrence of these being largely controlled by climate. The FAO closed forest survey covered all three types though thorn forest, which occurs in very dry and strongly seasonal tropical climates, has only limited global extent compared to the other two kinds. It is, however, extensive in some individual countries, for instance India.

The different closed forest formations are seldom distinguished in national statistics and are difficult to tell apart by remote sensing, especially in the wet season when seasonal forest trees are in leaf. The convenient term tropical moist forests has been coined to include both tropical rain and seasonal forests (Sommer, 1976).

Deforestation is the clearance of forest and the conversion of the land to some other usage, generally for agriculture. Alternatively, and often, primary forests are logged and left to recover, or they are felled and the land farmed for a few years and then allowed to redevelop forest. This process of forest alteration, sometimes called 'degradation' is less drastic than complete deforestation, because the land returns to a continuous tree cover and ecological functions more or less recover. However, species composition, original community structure and species' interdependencies are all disrupted to a greater or lesser extent.

Forest alteration was included with deforestation in the narrow sense in a major survey by the US National Academy of Science which used the term 'conversion' to cover all intensities of human interference from highly selective timber extraction to total transformation to agriculture (Myers, 1980 p.8). The same terminology was used in the recent repeat of the survey commissioned by Friends of the Earth (Myers, 1989, 1990). Because these two kinds of human impact are considered together the

Table 1.1. FAO survey on tropical forest clearance and logging. Data from Lanly (1982) for 76 countries which total 87% of all the tropical land area

	Area (10^2 km²/yr)	*Annual reduction (%)*
A Estimate for 1981–85		
1 Closed forest clearance (a)	750	0.62
Undisturbed closed forests logged but not cleared (b)	*c* 440	
Therefore, total 'altered' tropical closed forest (c = a + b)	1190	
2 Open tree formations (especially cerrado, chaco, of neotropics, savanna woodlands etc. of Africa) clearance (d)	380	0.52
3 Total clearance closed plus open forest (e = a + d)	1130	0.58
4 Total 'alteration' closed plus open forest (f = c + d)	1570	
B Guess for 1980–2000 (if linear extrapolation is justified)		
5 Closed forest cleared 1980–2000 (a × 20) 15000		

Myers' surveys give higher estimates of forest loss than the survey of FAO (Sayer and Whitmore, 1991).

Forest and non-forest are relatively easy to distinguish on remote-sensing images but it is difficult to monitor the various kinds of disturbance because it is very difficult to discriminate between primary (old-growth) logged and regenerating secondary forest.

The FAO survey was based on an evaluation of data compiled from existing maps of vegetation and landuse, forest inventories, satellite images and reports on colonization schemes and shifting agriculture, not (as has sometimes been stated) on figures provided by national forestry institutions (Lanly, 1991). For some countries (e.g. Burma, Laos: R. Persson pers.comm.) the data were scanty. Table 1.1 summarizes the final results. For closed forests, clearance (a) was distinguished from forests logged but not cleared (b). An allowance was made in (b) for illegal entry of farmers into those forests which should have been left to regrow. In open woodlands the progressive removal of trees, usually for fuel or fodder, cannot readily be monitored, and therefore the only category FAO recognized was total clearance (d). Various totals in Table 1.1 have come to be widely quoted: clearance of closed forests at $750 \times 10^2 km^2/yr$, (a); clearance of closed plus open forests at $1130 \times 10^2 km^2/yr$, (e). The figure for total 'alteration' (cleared + logged) of closed forests of 1190×10^2 km^2/yr is most closely comparable to the figures of Myers (1980, 1989, 1990) but an area of forest logged one year may be relogged or cleared later and thus counted twice (or more times) if annual totals are summed. Linear extrapolation of the annual rate of closed forest clearance from 1980 to 2000 gives a figure of $15\,000 \times 10^2$ km^2 lost by then, or 12.5% of the 1980 area.

FAO intends to issue a new report on the state of the world's tropical closed forests to a 1990 dateline, i.e. one decade later than its original report. It has already issued preliminary figures (FAO, 1990a,b) increasing its estimate of the annual forest clearance figure for 1981–5 from $1130 \times 10^2 km^2$ for 1981–5 (Table 1.1) to $1700 \times 10^2 km^2$ for 1981–90. The increase is due to underestimates of deforestation rates for some Asian countries in the original report, not, as has been widely stated, to an increase in global deforestation rates since 1980.

TROPICAL FOREST AREA ESTIMATES COMPARED

The new *Conservation Atlas of Tropical Forests* (Collins *et al.*, 1991) should permit comparison between the area of closed tropical forest in the Asia–Pacific region estimated to remain in 1980 by FAO (FAO, 1988) and the most recent subsequent estimate from which the maps in the atlas were prepared. The data are presented in Table 1.2. Taken at face value seven countries show a reduction in forest area during the 1980s and seven others an increase (for Cambodia there is no estimate since 1980). The countries at the top and bottom of Table 1.2 which show respectively a big decrease (India, Vietnam, Sri Lanka) and a big increase (Cambodia, Laos) in forest area during the 1980s in fact neatly epitomise the whole question of definitions which bedevils all investigations into tropical forest clearance. For example, FAO covers all closed tropical forests and in India this includes extensive thorn forest (see above); the latter is excluded from the conservation atlas. This also explains the big difference between the two figures for Sri Lanka. For Vietnam, Cambodia and Laos, which all show large differences, there are substantial problems in delimiting tropical moist forest in the atlas and subjective decisions were made (Whitmore, pers. comm.). The Philippine forests have been mapped as part of a German aid project (Forest Management Bureau, 1988) and the big reduction in forest area shown for the 1980s is probably real. For the other countries in Table 1.2 there is no way of ascertaining how much of the forest decrease shown for Fiji, Malaysia and Myanmar (Burma) is real. A decrease must have occurred during the 1980s in Indonesia, Bangladesh,

Table 1.2. Recent estimates of remaining area (in $10^2 km^2$) of tropical closed forest in Asia

	FAO data for 1980 (a)	*Most recent estimate* (b)		*a/b* (%)
India	5184	2283	1986	227
Vietnam	877	567	1987	155
Philippines	951	660	1988	144
Sri Lanka	166	123	1988	135
Fiji	81	70	1980s	116
Malaysia	2100	2005	1979–86	105
Burma (Myanmar)	3194	3119	1987	102
Indonesia	11389	11791	1985–89	97
Bangladesh	93	97	1981–86	95
Solomon Islands	242	256	1980s	95
Papua New Guinea	3423	3667	1975	93
Thailand	923	1069	1985	86
Brunei	32	47	1988	69
Cambodia	755	1133	1971	67
Laos	841	1246	1987	67

(a) FAO (1988) closed broadleaved, conifer and bamboo forests.
(b) Collins *et al.*, (1991) Table 11.1, moist forests.

Solomons, Papua New Guinea, Thailand and Brunei and the apparent increase in Table 1.2 must be an artefact attributable to differences in definition.

We conclude from this analysis that different estimates of tropical forest extent must be interpreted with extreme caution. The comparison between the FAO 1980 estimate and the new FAO estimate for 1990 currently in preparation will be of great value because both use the same definitions so will permit more accurate comparison. The difficulty of obtaining good basic data may still obscure the real state of affairs.

TROPICAL DEFORESTATION ESTIMATES COMPARED

Table 1.3 shows the 1980 FAO estimate of the annual clearance rate of tropical broadleaved closed forest (as corrected by FAO, 1988), plus other figures collated by Sayer and Whitmore (1991) and World Resources Institute (WRI) (1990) for these same countries, which collectively encompass over 90% of the original tropical moist forests (Myers, 1989).

Table 1.3. Recent estimates of annual rates of clearance of tropical closed forests for selected countries

	1981–85 rate of FAO (1988)		Sayer and Whitmore (1991)		WRI (1990)		Rate compared to FAO
	%	10^2km^2	year	$10^2km^2\ yr^{-1}$	year	10^2km^2	
New World							
Bolivia	0.2	8.7	–		–		–
Brazil	0.4	136	1987	350	1987	800	up
Columbia	1.8	82	1960–84	60	–		down
Costa Rica	4.0	6.5	1977–83	12.4	–		up
Ecuador	2.4	34	–		–		–
Mexico	1.8	47	–		–		–
Peru	0.4	26	1988	30	–		up
Venezuela	0.4	12.5	–		1977–83	12.4	same
Asia: continent							
Myanmar	0.3	10.15	c. 1980	60	1975–81	67.7	up
Cambodia	0.3	2.5	–		–		–
India	0.3	13.2	–		1975–82	150	up
Laos	1.3	10	–		–		–
Thailand[a]	3.0	37.9	–		1978–85	39.7	up
Vietnam	0.8	6	–		1976–81	17.3	up
Asia: insular							
Indonesia	0.5	60	c. 1985	100	1979–84	90	up
Malaysia	1.2	25.5	c. 1985	31	–		up
Papua New Guinea	0.1	2.2	–		–		–
Philippines	1.0	9.1	1969–88	21		–	up
			1988	13	1981–88	14.3	up
Africa: west							
Ivory Coast	6.5	29	–				–
Nigeria	5.0	30	–				–
Africa: central							
Cameroon	0.4	8	c. 1987	10			up
Congo	0.1	2.2	–				–
Gabon	0.1	1.5	c. 1987	15			up
Zaire	0.2	18	–				–
Africa: insular							
Madagascar	1.5	15	c. 1986	15–30			up

[a] includes open woodland.
Sources:
FAO (1988) Table 4, broadleaved closed forest; percentages are of original cover.
Sayer and Whitmore (1991), Table 3.
World Resources Institute (1990), Table 7.1.

Table 1.4. Deforestation in the Brazilian Amazon (Fearnside *et al.*, 1990)

Rate
1978 to 1989 annual loss 21 218 km^2 ± 10%
1989   loss in 1.05 years to August 1989 26 664 km^2

Area of Legal Amazonia deforested
January 1978   54 502 km^2, 1.1%[a]
April   1988 264 421 km^2, 5.4%[a]
August 1989 299 079 km^2, 6.1%[a]

---

[a] Add to these values 97 643 km^2 (2%) of old deforestation.

For all countries, present estimates of clearance are higher than the original FAO (1982) estimates. The only exceptions are Venezuela, where the figures are about the same, and Colombia. The problem of definitions discussed above for Table 1.2 is likely to affect Table 1.3 also, causing for example the huge apparent increase in deforestation in Myanmar and India. Brazil also shows an enormous increase in deforestation rate, which in this case reflects the very different figures being published for that country. These are partly due to the different forest formations and regions included in different studies and partly due to technical problems in the interpretation of remotely sensed images (see discussion in Fearnside *et al.*, 1990 and Sayer and Whitmore, 1991). Table 1.4 gives the figures for Brazil which Fearnside *et al.* (1991) argue are the most reliable.

The conclusion which can be drawn from Table 1.3 is that, despite problems in definitions, tropical forest clearance rates were underestimated in the early 1980s and have accelerated everywhere over the decade that has since elapsed.

SPECIES EXTINCTION

Species extinction rates will depend on the number of species present in (and, especially restricted to) particular areas of forest, the distribution of deforestation, and the extent of disturbance or degradation of those areas which remain forested. The distribution of species diversity in tropical moist forests is not only very patchy but also very poorly known, although knowledge is improving all the time (Whitmore and Prance, 1987). Sample plots of one hectare with over 200 species of tree ≥ 10 cm diameter have been known from Southeast Asia since the 1960s (Whitmore, 1984) and recently such a plot was enumerated in Peru (Gentry, 1988). Until recently, the maximum number of tree species recorded in one hectare in Africa was 73 (Whitmore and Sidiyasa, 1986), but 138 species have now been found on a plot of 0.64 ha at Korup, southwest Cameroon (Gentry, 1988).

Invertebrate diversity is least well known. In a famous extrapolation based on data from Panama, Erwin (1982) postulated that as many as 30 million arthropod species may exist. Re-examination of these data led Stork (1988) to conclude that there are between 20 million and 80 million species. This reasoned extrapolation is an advance on previous guesses at global species numbers and establishes a methodology for estimating global diversity (May, 1988). Most of these species, if they do indeed exist, live in tropical moist forests, but their distribution and status within the forest are very poorly known.

Estimates of plant and invertebrate extinctions are inevitably largely a matter of speculation. Consequences for species survival of the degradation, partial clearance and fragmentation of large forest areas are simply not known though biologists have begun to think about the problem (e.g Simberloff, chapter 4). It is more likely that there will have been massive extinction of locally endemic plants and animals (among which invertebrates are especially numerous) in those places where large areas of tropical moist forest have been completely destroyed. Three such areas are Mato Grosso and Para states in southeastern Brazilian Amazonia, where 9% of the forest (179 380 km^2) has been cleared for pasture (Fearnside *et al.*, 1990, Table 1); the Atlantic coastal rain forest of Brazil where about 95% has been cleared (Prance and Campbell, 1988); and the Philippines, once wholly forested and now reduced to 22% (Forest Management Bureau, 1988).

None of these regions had been well explored by biologists before their forests were destroyed and there is no way of knowing precisely how many of their species have become extinct. Even though we cannot estimate precisely the number of extinctions, such massive changes can be assumed to have led to substantial losses in these three regions, Many other tropical moist forests are likely to go the same way with many further extinctions. Even if extinctions cannot be demonstrated this should not be interpreted as meaning that the problem is not serious.

In some deforested areas some species may persist in tiny patches of forest left along gullies and rocky ridges, etc. Recent investigations described by Brown and Brown (chapter 6) into the Atlantic coast rain forests of Brazil show that many species can survive for a reasonable period in such fragmented forests. However, our knowledge of the processes underlying extinctions (Simberloff, chapter 4) suggests that the eventual extinction of these species may be inevitable.

The only groups for which reasonably comprehensive information on extinction is available are the vertebrates, especially mammals and birds. Here it is instructive to examine the occurrence of species in discrete areas of forest. Lists for national parks and other reserves provide such samples and can enable us to speculate on the implications for survival if all forest were to be cleared from all non-protected areas. The most comprehensive summary of vertebrate species occurrence in protected areas is given in the IUCN reviews for Africa and Indo-Malaya (MacKinnon and MacKinnon, 1986a,b), which demonstrate that existing protected areas, which cover 2–8% of the area of most tropical countries, include populations of nearly all bird and mammal species. In forested countries in Africa protected areas have been found to contain populations of 70–90% of the national bird faunas (Sayer and Stuart, 1989). These figures are an endorsement of the choice of areas for conservation, but their implications for long-term conservation of biological diversity need to be viewed with great caution. Small isolated populations are susceptible to extinction from inbreeding depression (though there is little direct evidence for the impact of this phenomenon, Simberloff, 1988) and a variety of random causes. Area alone is a poor predictor of the conservation value of forest fragments (Zimmermann and Bierregaard, 1986). In at least some situations, birds may persist for numerous generations at very low population levels (Dowsctt–Lemaire, 1983). We interpret these data as demonstrating that massive extinction of vertebrates is unlikely during the next few decades, but that the the long-term survival of many species which now only persist as isolated populations in relatively small protected areas is in doubt unless the habitat available to them can be increased (see Simberloff, chapter 4).

CONSERVATION PRIORITIES

The above arguments lead to the conclusion that to optimize the application of resources, one should at least embark upon a 'minimum programme' to ensure the effective conservation management of those forest sites in the humid tropics that are known to support high levels of species diversity and endemism. Many of these are already legally gazetted parks and reserves. The viability of this strategy clearly depends upon the maintenance of reasonable environmental stability in the surrounding lands. It does not in any way diminish the imperative for non-protected forest lands to be brought under management systems that will provide other environmental services, such as the regulation of climate and water supply. If forests surrounding protected areas approach the structure and species composition of the natural forest they also allow extensions of the range and population size of many species. Such a scenario will favour the maintenance of more species than would survive if the protected areas were surrounded by less diverse systems. The optimal form of land use outside parks and reserves is near-natural forests under sustainable management for timber or non-timber products (Peters *et al.*, 1989; Sayer, 1991).

CONCLUSIONS

Information on the distribution of vertebrate diversity shows clearly that the relationship between forest loss and species loss is not arithmetic. To extrapolate upon such a relationship presents an excessively pessimistic view.

Extinction estimates should not be based upon extrapolation of forest loss towards a zero base. All countries have some permanent forest estate or inaccessible forests which are reasonably secure. In many cases these are part of a national protected area system and are located so as to cover a wide range of representative and biologically rich sites.

Studies of the flora and fauna of protected areas could enable us to determine the number of species which they contain, and which could be protected from extinction if the integrity and stability of the areas could be secured. In so far as the total number of species in a taxon is known, this would enable us to formulate some idea of global extinction rates. This approach is likely to provide a more useful management tool for determining priorities for conservation than theoretical estimates based upon generalizations from inadequate data. However, the changes in temperature and rainfall associated with even the most conservative predictions of climate change will have severe ecological impacts on protected areas.

The greatest causes for concern are twofold. First, forest loss is in general greatest in those places where least forest remains, for example West Africa and Madagascar. The incremental extinction resulting from further forest loss in places where forest is already fragmented and of limited extent is likely to exceed that following the loss of an equivalent area in one of the major forest blocks. Second, the reduction of forest area to fragments reduces numbers of individuals and perhaps loss of some ecotypes so that even if a species survives it has lost much of its genetic diversity. This is less obvious than complete loss of the species, but is likely to be an insidious but widespread consequence of current tropical deforestation (see Heywood and Stuart, chapter 5).

SUMMARY

Tropical moist forests are extremely rich in animal and plant species. They are now being altered or cleared more rapidly than at any time in the past. FAO (1988) has recently reported that deforestation rates during the 1980s were higher than their earlier published accounts had suggested. Comparisons of estimates of forest loss are made difficult by inconsistent use of definitions of forest cover types and of deforestation. Even allowing for this some recent assessments have shown accelerating deforestation rates in the 1980s for many individual countries.

The total number of species in tropical moist forests is not known. The pattern of likely species extinctions depends on the amount and spatial arrangement of forest altered or destroyed. A better understanding of the processes of deforestation and their implications for species extinctions is essential in order to set priorities for conservation. Loss of genetic diversity within species is even less easy to detect than species extinction but is undoubtedly very prevalent and is an issue of major concern for conservation.

ACKNOWLEDGEMENTS

This chapter advances the argument of Sayer and Whitmore (1991) in the light of newly available information, notably the contributions to this book.

REFERENCES

Collins, N.M., Sayer, J.A. and Whitmore, T.C. (eds) (1991) *The Conservation Atlas of Tropical Forests: Asia and the Pacific*. Macmillan, London.

Dowsett-Lemaire, F. (1983) Ecological and territorial requirements of montane forest birds on the Nyika plateau in south central Africa. *Gerfaut*, 73, 3345–78.

Ehrlich, P.R. and Ehrlich, A.H. (1981) *Extinctions*. Random House, New York.

Erwin, T.L. (1982) Tropical forests: their richness in Coleoptera and other arthropod species. *The Coleopterists' Bulletin*, 36, 74–5.

FAO (1981) *Tropical Forest Resources Assessment Project*. 4 vols, FAO, Rome.

FAO (1988) *An Interim Report on the State of Forest Resources in the Developing Countries*. Forest Resources Division, Forest Dept. FO:MISC/88/7. FAO, Rome.

FAO (1990a) *TFAP Independent Review Report 1990*, Appendix 3, section 3.3 (pp. 62–3).

FAO (1990b) *Interim Report on Forest Resources Assessment 1990 Project*. Committee on Forestry 10th Session. COFO-90/8(a).

Fearnside, P.M., Tardin, A.T. and Meira Filho, L.G. (1990) *Deforestation Rate in Brazilian Amazonia*. National Secretariat of Science and Technology, 8pp. typescript.

Forest Management Bureau (1988) *Natural Forest Resources of the Philippines*. Philippine–German Forest Resources Inventory Project. (pp. 62 mimeo).

Gentry, A.H. (1988) Tree species richness of upper Amazonian forests. *Proceedings National Academy of Sciences USA*, 85, 156–9.

Lanly, J.P. (1982) Tropical forest resources. *FAO Forestry Paper*, 30.

Lanly, J.P. (1992) The status of tropical forests, in *Tropical Forests* (ed. A.E. Lugo), Springer, New York (in press).

Lugo, A.E. (1988) Estimating reductions in the diversity of tropical forest species, in *Biodiversity* (eds E.O. Wilson, and F.M. Peter), National Academy Press, Washington DC, pp. 58–70.

MacKinnon, J. and MacKinnon, K.S. (1986a) *Review of the Protected Areas System in the Afrotropical Realm.* IUCN, Gland and Cambridge.

MacKinnon, J. and MacKinnon, K.S. (1986b) *Review of the Protected Areas System in the Indo-Malayan Realm.* IUCN, Gland and Cambridge.

May, R.M. (1988) How many species are there on earth? *Science*, **241**, 1441–9.

Myers, N. (1980) *Conversion of Tropical Moist Forests.* National Academy of Sciences, Washington.

Myers, N. (1983) *A Wealth of Wild Species.* Westview, Boulder.

Myers, N. (1989) *Deforestation Rates in Tropical Countries and Their Climatic Implications.* Friends of the Earth, London.

Myers, N. (1990) Tropical forests, in *Global Warming – the Greenpeace Report* (ed. J. Leggatt), Oxford University Press, Oxford.

Peters, C.M., Gentry, A.H. and Mendelsohn, R.O. (1989) Valuation of an Amazonian rain forest. *Nature*, **339**, 655–6.

Prance, G.T. and Campbell, D.G. (1988) The present state of tropical floristics. *Taxon*, **37**, 519–48.

Raven, P.H. (1987) The Scope of the Plant Conservation Problem World-wide, in *Botanic Gardens and the World Conservation Strategy* (eds. D. Bramwell, O. Hamann, V. Heywood and H. Synge), Academic Press, London, pp. 19–29.

Sayer, J.A. (1991) *Rainforest Buffer Zones: Guidelines for Protected Area Managers.* IUCN Gland, Switzerland and Cambridge, UK.

Sayer, J.A. and Stuart, S.N. (1989) Biological diversity and tropical forests. *Environmental Conservation*, **15**, 193–4.

Sayer, J.A. and Whitmore, T.C. (1991) Tropical moist forests: destruction and species extinction. *Biological Conservation*, **55**, 199–214.

Schimper, A.F.W. (1903) *Plant-Geography upon a Physiological Basis* (translated W.R. Fisher, P. Groom and I.B. Balfour), Oxford University Press, Oxford.

Simberloff, D. (1986) Are we on the verge of mass extinction in tropical rain forests?, in *Dynamics of Extinction* (ed. D.K. Elliott), Wiley, New York.

Simberloff, D. (1988) The contribution of population and community biology to conservation science. *Annual Review of Ecology and Systematics*, **19**, 473–512.

Simon, J.L. (1986) Disappearing species, deforestation and data. *New Scientist*, May 15.

Sommer, A. (1976) Attempt at an assessment of the world's tropical forests. *Unasylva*, **28** (112/113), 5–25.

Stork, N.E. (1988) Insect diversity: facts, fiction and speculation. *Biological Journal of the Linnean Society*, **35**, 321–7.

Whitmore, T.C. (1984) *Tropical Rain Forests of the Far East*, 2nd edn, Clarendon Press, Oxford.

Whitmore, T.C. and Prance, G.T. (eds) (1987) *Biogeography and Quaternary History in Tropical America.* Clarendon Press, Oxford.

Whitmore, T.C. and Sidiyasa, K. (1986) Composition and structure of a lowland rain forest at Toraut, northern Sulawesi. *Kew Bulletin*, **41**, 747–56.

Williams, M. (1989) Deforestation: past and present. *Progress in Human Geography*, 176–208.

Wolf, E.C. (1987) On the brink of extinction: conserving the diversity of life. *Worldwatch Paper*, 78.

World Resources Institute (1990) *World resources 1990–91*. Oxford University Press, Oxford.

Zimmermann, B.L. and Bierregaard, R.O. (1986) Relevance of the equilibrium theory of island biogeography and species–area relations to conservation with a case from Amazonia. *Journal of Biogeography*, 13, 133–43.

– 2

Species conservation in managed tropical forests

A.D. JOHNS

INTRODUCTION

Within the last few years there has been an increasing awareness of the need for development of tropical rain forest management systems. Very little of the world's rain forest area can be considered to be under effective management at present: perhaps less than 1% (Poore, 1989). This is despite a current level of exploitation of *c*. 50 000 km^2 logged or re-logged annually (Sayer and Whitmore, 1990). There is little doubt that much of the remaining tropical forests will be exploited for timber during the next few decades. It is of international concern that such exploitation should take place in a controlled manner; that is, that sustainable harvesting systems replace current exploitation.

Silvicultural systems have been devised for some tropical rain forests, and enough is known about their necessary features and about local ecological characteristics to devise them for most others (Whitmore, 1990). The essence of a good forest management system is that the production of timber and other products be sustainable indefinitely. The description 'sustainable' has been applied in a number of ways, often unjustifiably. In this chapter, the foresters' definition is adopted, namely:

> To harvest forest in such a way that provides a regular yield of forest produce without destroying or radically altering the composition and structure of the forest as a whole . . .

This description (adapted from Wyatt-Smith, 1987a), and any others that have so far been produced, inevitably includes an indeterminate qualifier: in the above case it is the word 'radically'. Means whereby sustainability can be further defined are examined later.

The definition of Wyatt-Smith goes on to address a second major issue, that of management to enhance timber production, namely:

... Thus it is more than maintaining the rate of production at the low level of natural forest, which merely replenishes natural mortality. It is used in the sense of enhanced production through silvicultural practices, while conserving the protective role of the vegetation and the genetic pool of all species other than those regarded as weed species which compete with, and suppress, favoured timber trees.

This is problematic, since sustainable operations will be required to mimic natural ecosystem processes. Some enrichment may be possible, however, without causing a divergence of ecosystem processes.

Large areas of logged-over forest will remain in the tropics through the foreseeable future. If these are managed so as to support the long-term production of diverse natural timbers, they should also support the long-term conservation of at least some of the animal and plant species of undisturbed forest. As such, these areas become important components of national biological diversity conservation programmes. For various reasons, such as their role in the pollination and seed dispersal of many forest trees, wildlife will, in any case, be an integral part of a successful forest management operation. It is important to co-ordinate the work of foresters and conservationists in attempts to maintain both timber (and non-timber) products and the species diversity of tropical countries.

At the present time, much logged forest is unmanaged. Almost everywhere sustainable utilization is an ideal rather than common practice. Very few logging operations can justifiably claim to be practising sustainable forest management as defined above. In most areas, however, the ideal should be attainable. It will require at least a partial re-think of current extraction practices, but should be possible within the economic constraints faced by the tropical timber industry.

The issues

Timber management

Global consumption of wood products is currently running at around 3000 million m^3 annually (Whitmore, 1990). About 55% of this is fuelwood; of the 45% industrial forest products, two thirds is composed of pulp and the remaining one-third of timber.

Timber exports from tropical countries are worth around US$ 7500 million annually, ranking fifth among non-oil exports. Almost 60% of tropical forest timber products traded originate in three South-east Asian countries: Malaysia, Indonesia and the Philippines (the last of lesser important since 1986). Logging in the dipterocarp forests of South-east Asia is generally less selective than in Africa and the Neotropics, because the former contain a much higher standing volume of merchantable timber. However, dipterocarp forests are mostly no better managed than those elsewhere (Wyatt-Smith, 1987a).

There are a few places where selective logging and silvicultural treatment have perhaps been carried out successfully. In some cases, repeat cuts have been obtained after 30–70 years without any obvious (subjectively determined) changes in the character of the forest concerned, other than the targeted increase in the proportion of exploitable species. Examples of apparently successful management systems have been reported from Uganda (Plumptre and Earl, 1984), Surinam (De Graaf, 1986; Jonkers, 1987), and Burma (IUCN, 1987). Some success has been claimed in South-east Asian dipterocarp forests (Wyatt-Smith, 1987a), but this is not true of most timber operations (Tang, 1987). Even in South-east Asian forest it has been shown that changes could be made which would improve the efficiency of logging operations (Marn and Jonkers, 1981; Marn, 1982). It is certain that most tropical moist forests could be managed as a truly renewable resource, if human intervention operated within the inherent limits of the natural cycle of growth and decay found in all forests (Whitmore, 1990). Unfortunately, even those management techniques that have been developed are rarely put into practice. Logging is most often controlled by entrepreneurs, to whom short-term profits are of prime importance, rather than by foresters, whose duty is to the long-term maintenance of the resource.

The conservation of biological diversity

Very little of the remaining tropical forest area is currently legally protected as national parks or equivalent inviolate reserves. Even if existing proposals are implemented, these will not exceed 7% of remaining tropical forests. The amount of forest that is *de facto* physically protected from intrusion by hunters or settlers, or excluded from oil and mineral prospecting, hydroelectric schemes and highway developments is very much less.

In many tropical countries, at least 70% of the resident vertebrate species are commonly reported to be dependent upon closed forest. The corresponding figure for invertebrates is probably much higher (Erwin, 1988). Furthermore, local endemism is very high in many tropical regions. These facts have led to the widely held belief that the area of primary forest protected in national parks and equivalent reserves is insufficient to maintain biological diversity in most regions.

Two basic problems have faced wildlife preservation initiatives: a shortage of undisturbed areas and the degradation of what little has already been set aside. In the first case, financial problems and land use trends often preclude the preservation of large areas of forests not producing revenue and which require capital in order to be maintained. In the second case, few reserves can be considered inviolate, particularly in regions with rapidly increasing rural populations.

Production forests are rarely considered in overall conservation strategies. It is often thought that separate areas are required for the two 'opposing' goals. For example, Wyatt-Smith (1987b) considers that, in Malaysia:

> The managed lowlands ... will not resemble the primary forests in either structure or species composition and diversity ... Conservation and present-day timber exploitation are no longer compatible and require independent land allocation.

This viewpoint has been disputed (e.g. Johns, 1985, 1989; Whitmore, 1990) on the grounds that a sustainable forest management system should maintain environmental and ecological parameters in a near natural state. Plantation forestry may be necessary in some areas to maintain national timber production at current high levels (e.g. in Sabah, Tang, 1987), but this should be considered as an alternative use for land already deforested rather than a prospect for natural forest management. Sustainable forest management necessarily includes measures for conservation, and these must be employed at an early stage of forest development (before the area concerned becomes subjected to unnecessarily destructive forestry techniques or is invaded by settlers). It may be possible to regenerate natural productive forest from damaged remnants, but this will be a long-term process and an inefficient use of the resource. The sooner that ecologically sustainable forest management strategies are adopted, the more efficiently the resource will be utilized in the long term.

Rationale

The tropical timber industry faces a number of problems of its own, including inefficient species utilization, an inadequate infrastructure, and so on. These constraints need to be appreciated in planning integrated sustainable management strategies. The natural regeneration potential of the forest sets limits to the activities of foresters corresponding to the resilience of the ecosystem. Forest recovery following logging is dependent upon a variety of factors, of which the most important include how much of the stock of seedlings and saplings survive logging, the presence of natural pollinating and dispersal agents, and the maintenance of soil fertility. It is necessary to maintain a balanced ecosystem because only a balanced system can be sustainable. High species diversity will of necessity be a part of a balanced system.

Managed tropical forests may not contain or regain a comparable complement of plant and animal species unless there is provision within management plans for the preservation of unlogged areas. Sources of colonists may be required. Even poorly managed logged forests have been shown to retain remarkably high numbers of animal species, however (Johns, 1985). This occurs without the preservation of very large unlogged areas within timber concessions. Preservation of only around 5% of an intensively logged area in Sabah, for example, has been shown to preserve populations of almost all bird species (Johns, 1989). It requires very little effort or economic sacrifice on the part of the concession owner to assist the retention of vertebrate species richness through the provision of refuges. It requires more effort to assist the retention of species richness by adoption of appropriate harvesting strategies.

There is regrettably little information available concerning the responses of many species groups, however. Current views are biased by the responses to larger animals, which are only a part of the total biological diversity of the forests. While admitting the limitations of current knowledge, the continuing loss of undisturbed forest can only mean that managed areas will be a principal hope for forest species of all types.

This chapter evaluates current data concerning timber and wildlife resource management in tropical rain forests to examine the extent to which forest species may exist alongside current management procedures. Approaches are suggested that will enable production and conservation objectives to be reconciled.

BACKGROUND

Logging industries

Exploitation of tropical timber is increasingly a highly mechanized capital-intensive industry, particularly in South-east Asia, where logging is primarily geared towards the export market. Labour-intensive hand-sawing is still found in some areas of Africa and the Neotropics, but is generally directed towards supplying the domestic market.

An export-oriented logging company will generally require a concession with a minimum yield of 100 000 m³/year and a minimum 15-year lease to support the construction of a modern, efficient logging unit and sawmill. Many companies operate with far less than this, transporting logs to already established local sawmills or, if legal, exporting round-logs. Other companies work as contractors, helping larger concerns to meet schedules as necessary. A logging contractor, or a company with a short-term lease, will be concerned only with a single cut and will not be motivated to minimize environmental damage. Some tropical countries have placed legal restrictions on damage levels, but 'unnecessary damage' has proven difficult to define in practice. Lack of financial interest in the future crop is probably the main reason for the excessive damage levels typical of short-term extraction operations.

Timber resources are being depleted rapidly. The short-term nature of timber operations is incurring unacceptable environmental costs for many countries. In Nigeria, for example, all forested land has now been put under the control of the Forest Department, which aims to manage productive forests on a scientifically-determined sustainable basis. Unfortunately, this occurred after the forests had been severely depleted and sustainable use is unlikely to be achieved: the country is now a net importer of timber products. Some countries with large areas of forest remaining, such as Papua New Guinea, require logging companies to replant logged land to defined standards, or to pay a Government-sponsored agency to do so. Again, however, this rarely works in practice because the companies often find it more expedient to default and pay any resulting fines. There is an increasing trend towards delimiting and maintaining productive forest as a financial resource in perpetuity, but there are many problems to be overcome before this becomes standard practice.

Current logging practices

Commercial logging in tropical rain forests can take a number of forms, almost all of which involve the removal of selected trees rather than the clear-felling of whole stands. The only exceptions are three operations (Colombia, Papua New Guinea, Sabah) where forests are clear-felled for wood chips. This is different from temperate forestry operations where clear-felling is much more common.

Timber harvesting is basically an exercise in transport at minimum cost, since the product is large and heavy. The distances over which it is possible to transport timber vary according to its value. In Amazonia, up to 140 species may be logged in the eastern forests accessible both to local markets and to the populated regions of southern Brazil (Uhl and Viera, 1989), whereas as few as two or three species may be cut in the isolated western regions (Johns, 1988a). On a global scale, the major sources of tropical hardwoods in West Africa and South-east Asia are a reflection of the preponderance in these areas of light- and medium-density hardwoods of high utility value. The greater proportion of high-density hardwoods over much of Amazonia will largely preclude similarly intensive harvesting for export (Whitmore and Silva, 1990).

Non-intensive timber harvesting

Most inhabitants of tropical rain forest cut down trees for domestic uses, such as building houses or canoes. This is primarily small-scale, but it can involve the over-exploitation of certain tree species (e.g. Marsh, 1980). In some parts of the world, even commercially-oriented logging is still a family operation using minimal technology. In the western Amazonian *várzea* and *igapó* swamp forests, for example, loggers cut trees when water levels are low and float them to sawmills during the flooded season, all without a need for large and expensive machines and with comparatively little damage to the forests (although as timber values increase, stocks can become severely depleted). Similarly, in parts of Africa hand-cutting and pit-sawing of timber trees has been suggested to be perhaps a more efficient use of resources (given low labour costs) than would be the case if machinery and high-capacity sawmills were developed (Struhsaker, 1987).

Intensive timber harvesting

The development of tropical rain forest timber resources rises through the scale of traditional use, commercial use of a few species, intensive use of many species, and finally (theoretically) complete utilization. In Queensland, for example, only one species, Red cedar, *Toona australis*, was used by the early settlers, 10 species were used by 1900, 30 species by 1930, and over 100 species by 1945 (Freezailah, 1984). This is a function of the growing scarcity of prime species and thus the greater acceptance of alternatives and later of small-sized species (Leslie, 1987). The number of species used is increasing in many parts of the world.

Intensive timber harvesting operations are characterized by the use of heavy hauling and transporting machinery, and, usually, building of appropriate processing and marketing infrastructures. Operations are typically capital- rather than labour-intensive. They arose mainly since 1950 and resulted in a 13-fold increase in the use of tropical hardwoods by the main importing countries (Whitmore, 1990). At this time, there were important technological advances, particularly the introduction of the one-man chain saw and better road-building and log-hauling vehicles. The move to capital-intensive operations required a higher extraction rate to cover the initial costs, and resulted in the development of new silvicultural systems such as the Malayan Uniform System (MUS) (Wyatt-Smith, 1963; Whitmore, 1984).

Capital-intensive operations take a variety of forms, which may be broadly divided into monocyclic and polycyclic systems. Monocyclic systems involve the removal of all merchantable trees in a single operation, usually with intensive post-felling treatments to promote the growth of desirable species. A second crop is available after about 70 years. Polycyclic systems practice a lower initial felling level, designed to limit damage to advanced regeneration of commercial species which become a viable second crop after only 20–30 years (Whitmore, 1990).

Recent years have seen a re-evaluation of ways of enhancing yields. In Neotropical and African forests, where polycyclic systems are operated and yields have been low (averaging 8.4 and 13.5 m^3/ha respectively in the early 1980s) (Freezailah, 1984) investigations have generally been towards increased species utilization (Asabere, 1987). In dipterocarp forests of South-east Asia the stimulus to change has been the long rotation times of monocyclic systems, such as the MUS, coupled with rapidly diminishing areas of unlogged forest. Utilization is already very high (generally more than 50 m^3/ha and up to 110 m^3/ha in Sabah) and further increases in extraction levels would certainly be detrimental to the regenerative properties of the forest. Here, polycyclic systems, which aim to encourage the growth of pole-sized trees of commercial species to form the next crop, have been developed. Such systems often imply replanting of damaged land areas to speed up regeneration of the timber crop. Whitmore (1984, 1990) considers these systems as workable, providing conscientious forest management is practised. Whitmore pointed out, however, that no actual evidence for the long-term success of such operations exists.

THE TREE COMMUNITY

Introduction

The overall impact of logging operations is dependent on two main factors, the number of trees removed and the care taken in so doing. A third factor is of considerable importance to forest-dwelling animals, namely the extent to which timber trees are important as food sources for particular species. Both the vegetation and animals are subsequently influenced by a fourth factor: the extent and type of post-felling silvicultural treatments applied to the logged-over forest.

Although tropical rain forests contain a very large number of tree species, only a few may be acceptable to the timber trade. In Peninsular Malaysia, for example, there are around 2500 tree species, of which 700 reach a size large enough to be usable (basal girth of > 1.35 m; Whitmore,

1984). Of these species, 402 are considered commercial, but even at the peak of timber exporting from the country, less than half (30 species and species groups) were exported in significant quantities. The degree of selection is even more pronounced in other tropical forest areas.

Damage levels

Causes of damage

In Malaysian dipterocarp forests dominated by consociations of *Dryobalanops aromatica* or *Shorea curtisii* the number of trees felled may reach 72/ha, equivalent to a stump every 12 m and causing almost total destruction of the forest canopy, but this is the extreme and 14/ha giving a stump every 25.5 m is the average (Whitmore, 1984). On the other hand, most Amazonian *terra firme* forests yield only 3–5 trees/ha (Johns, 1988a) and some African forests as few as 1.1 trees/ha (Bullock, 1980).

Removal of marketable trees is, however, only a minor consequence of logging. Cut trees are generally large emergents and their felling causes considerable damage to other layers of the forest (an emergent tree of > 2.5 m girth will destroy around 0.02 ha on falling: Dawkins, 1959). Construction of main access roads for removal of timber is commonly accompanied by the clearing of 20–30 m wide strips through the forest (this 'daylighting' is designed to aid drying out of the roads after rain); these main roads and their associated loading or landing areas occupy from 6 to 20% of the forest (Hamzah, 1978; Malvas, 1987). Skidroads, constructed by tractors moving between the cut trees and the main access roads, average 4 m in width and may total 27 km in length per km^2 of productive forest logged (figure from Peninsular Malaysia, Whitmore, 1984, where extraction averages about 50 m^3/ha). These cause more damage than anything else. Kartawinata (1978) estimates that 30–40% of a logged forest in Indonesia may be left bare of vegetation as a result of roading and dragging activities. Causes of tree loss at a number of sites underline the importance of damage to residuals (Table 2.1).

Insufficient information is available concerning the extent to which trees surviving logging are sufficiently damaged that they die some time later, for example through fungal infection of wounds, but some kind of damage to 40% of trees in the remnant stand is common (Abdulhadi *et al.*, 1981). Trees may also die as a result of insolation or water stress, or perhaps through a loss of symbiotic mycorrhizae as a result of these. Tree loss through windthrow is also much higher in logged forests. The uneven nature of the canopy increases wind turbulence, to which shallow rooted, remnant old trees are particularly susceptible.

A single violent storm in an Amazonian study site caused 19 treefalls/km^2 in an 11-year-old logged forest compared to 2.2 treefalls/km^2 in

Table 2.1. Causes of tree mortality during logging

	% Loss of trees (> 30 cm girth)				
	Ponta da Castanha, Brazil (1)	Nigeria (2)	S. Tekam Malaysia (3)	S. Pagai, Malaysia (4)	G. Tebu, Malaysia (5)
Killed					
Timber trees	0.6	1	3	8	10
Destroyed during construction of access roads and landing sites, etc.	60.4	25	8	46	55
Destroyed during felling and dragging			39		
Remaining	39	74	49	46	35

(1) Johns (1986a).
(2) Redwood (1960).
(3) Johns (1986b).
(4) Whitten *et al.* (1984).
(5) Burgess (1971).

adjacent unlogged forest, with an average of 4.7 trees of > 30 cm girth destroyed per treefall (Johns, 1986a).

Can damage levels be reduced?

The extent of initial change following logging may be strongly correlated with the care taken during felling operations. A study in Sarawak showed that felling trees in the direction which caused least damage, careful siting of skidroads and the restriction of tractors to them reduced damage levels by half without increasing costs (Marn and Jonkers, 1981; Marn, 1982).

A correlation has been suggested between the basal area extracted and damage levels (Nicholson, 1958). This assumes, among other things, an equal degree of care, or more likely carelessness, in all forests. In practice, there is less scope for carelessness in forests logged without using heavy machinery, and damage levels will be lower as a result. It is clear from a global analysis that damage levels realized by mechanized logging operations far exceed the minimum possible (calculated by mathematical simulation: Fig. 2.1). The Sarawak example (Marn and Jonkers, 1981; Marn, 1982) suggests that damage levels could be brought closer to the theoretical minimum if there was the political will to do so.

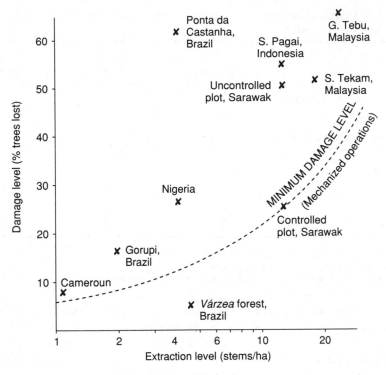

Figure 2.1. Damage levels under different logging intensities. Minimum damage levels are calculated from a mathematical simulation. Calculated minimum damage levels (untransposed data) are described by the line $y = 1.24x + 7.5$. From A.D. Johns (unpublished data).

Tree species richness

Where small areas are considered, tree species richness per unit area is generally slightly reduced as a result of logging (Table 2.2). Loss may be influenced by the basal area reduction during logging. This is, however, somewhat misleading. Through the forest as a whole, there may not be a reduction in species richness or diversity (Fig. 2.2). Results from Queensland in fact suggest that logging reverses a natural loss of diversity that occurs as a forest matures after disturbance and pioneer species die out (Nicholson *et al.*, 1988). However, logged forests may not always regain all species lost during logging and their species composition may be somewhat different.

During an intensive logging operation, only a very small proportion of trees destroyed are deliberately targeted. The result is that logging damage

Table 2.2. Comparative measures of tree species diversity in unlogged and logged forest in Uganda

	Unlogged	Lightly logged[a]	Heavily logged[b]
$\bar{x}$ species found in 100 stem sample	25.6	23.9	18.2
$\bar{x}$ species enumerated in 5 × 50 m plots	23.3	22.7	14.3
Equitability Hills eveness measure	0.62	0.53	0.50
Diversity Shannon–Wiener index	2.76	2.48	2.21

[a] 15% of trees destroyed 10 years previously
[b] 60% of trees destroyed 10 years previously
  Data from Skorupa (1986); a mechanical system was employed.

is spread over all tree taxa, with damage essentially random (Johns, 1988b). Given a random destruction, all rare species will be susceptible to depletion, but this will be particularly pronounced for rare species that are also highly valued timbers. The latter case will be true even under light logging schedules: a species such as Brazilwood, *Caesalpinia echinata*, has already been eradicated through most is its former range in the Amazon basin.

Following logging, the diversity of the forest may be enhanced by the rapid germination of pioneer tree species (which include many Euphorbiaceae in South-east Asia; *Cecropia* spp. and *Vismia* spp. and some legumes, such as *Inga*, in the Neotropics; *Albizia*, *Trema* and *Musanga* in Africa). In intensively logged forest, these colonizing species can be predominant. In Peninsular Malaysia, for example, a cohort of regenerating euphorbs (mostly *Macaranga* spp.) occupied 38% of a sample of trees > 30 cm girth in 5–6 year-old logged forest (Johns, 1988b). In such cases, changes in species composition in regenerating forest are considerable.

Trees as food sources

Animals in recently logged forest may face three problems concerning food source trees: (1) fewer trees, (2) a different spatial distribution of trees, and (3) different patterns of fruit and leaf production.

It should be recalled that the effects of most logging operations are

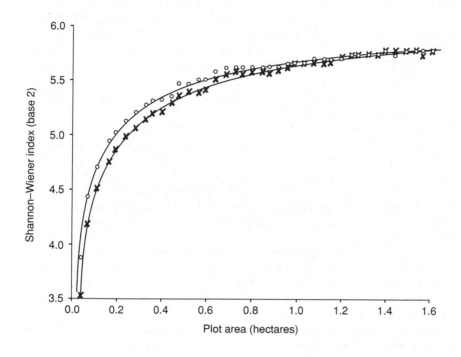

Figure 2.2. Effect of logging and plot size on tree species diversity (Queensland). $\circ$, before logging. $Hs = 6.2242 \exp (-0.09798 \text{ area}^{-0.48991})$; x, after logging. $Hs = 6.3864 \exp (-0.12923 \text{ area}^{-0.47163})$. From Nicholson *et al.* (1988).

broadly random, but that a highly desired timber species, particularly if rare in the forest, may be eradicated. If an animal is specialized to exploit such a species, then it may be eradicated also. In a slightly seasonal rain forest in Peruvian Amazonia, Terborgh (1983) estimated that as few as 1% of the trees may support the bulk of the frugivorous animal populations for several months each year. If logging were to remove the bulk of these trees the consequences would be serious. The effect of food tree eradication may be moderated, however, if the species are subsequently replanted for re-cropping at a later date (as is the case with *Calophyllum inophyllum* on Zanzibar Island (F.Omari, pers. comm.).

Food resources for frugivores tend to be widely dispersed in the forest, those for folivores often less so. The distribution of trees becomes more patchy following intensive logging (Johns, 1988b), which may necessitate changes in ranging and foraging behaviour among animals. Species that cannot readily adapt will be placed at a competitive disadvantage.

Chivers (1972) suggested that the initial loss during logging of trees used as food sources by wildlife may be buffered by greater levels of fruiting and new leaf production stimulated by the opening-up of the canopy. Johns (1988b) later showed that increased light availability does indeed result in increased vegetative activity among surviving trees, and may well increase the percentage of trees fruiting each month. The overall effect in recently logged forest was that the total availability of trees bearing new leaves did not decrease, and that the total availability of fruit might not, despite a drastic reduction in the numbers of trees per unit area.

In older logged forests the initial burst of vegetative and reproductive activity is somewhat reduced. Eleven-year-old logged forest in Brazil showed equal levels of new leaf production to primary forest (although there were differences in seasonality of leafing, corresponding to the differences in species composition) but a significantly lower level of fruiting (Fig. 2.3).

The extent to which pioneer tree species are used as food sources by animals varies considerably between geographical regions. In dipterocarp forests, neither the main climax timber trees nor the common pioneers are used frequently by animals as food resources (except as seedlings where they may be eaten by terrestrial browsing mammals). In Africa and the Neotropics, colonizing trees are important food sources for many species of vertebrates. The extent to which pioneers support primary forest invertebrates is unknown, but since many folivorous insects are specialists, it is unlikely that the regenerating forest will support a similar fauna.

The full diversity of food sources for forest animals is unlikely to be regained until regenerating seedlings of the primary forest trees reach reproductive maturity (which may take more than 20 years in a dipterocarp forest). Even after 50 years, regeneration of upper canopy epiphytes and other species dependent upon large mature trees (such as strangling figs, *Ficus*) may be incomplete. In general terms, however, a sufficient diversity and quantity of food resources are available following logging to support the majority of forest vertebrates, as is reflected in available population statistics.

THE FAUNA

Introduction

Rain forest faunas are exceedingly complex and diverse. Animals exist within a habitat mosaic, each patch of which is occupied by a somewhat different array of organisms, some transient and some resident. The degree

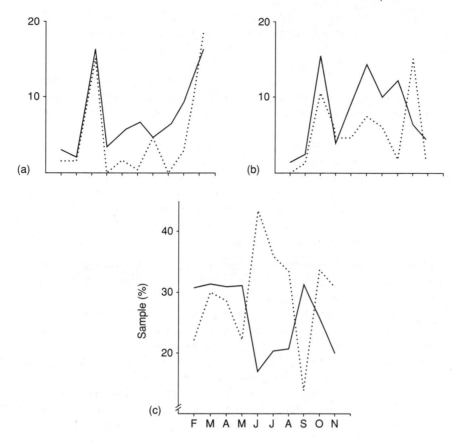

Figure 2.3. (a) Fruit, (b) flowers and (c) new leaf production in unlogged and logged forest, western Amazonia. Solid line, unlogged; dotted line, logged. Comparing matched samples of 66 trees there are no significant differences in vegetative activity, but logged forest showed significantly less fruiting and flowering activity (Mann–Whitney U tests: $P < 0.05$). From Johns (1986a).

of transience may be extremely high (e.g. Karr and Freemark, 1983) but depends on the mobility of the animals concerned. The extent of microhabitat specialization varies considerably, being most evident among small folivores (such as lepidopteran larvae which may rely upon certain parts of single species of foodplant) and least evident among large-bodied wide-ranging species with generalized diets.

Responses to the changed conditions of habitat and food supply brought about by logging can also be very varied. The extend of dietary specialization may be important in that animals specialized to exploit food sources that are less common following logging should, logically, also become less common. A reliance upon particular features of microclimate, or physical environmental characteristics that are changed by logging, may also have a deleterious effect. On the other hand, theorists have argued that a complex food web should show considerable resilience to environmental fluctuations since flexibility of connectance is an important feature of such systems: disturbances affecting certain component species should be absorbed by a redirection of community relationships (MacArthur, 1957).

In order to determine possibilities for the maintenance of biological diversity in managed forests it is axiomatic that it is more the form of the management operation that is under consideration than responses of individual species to defined disturbances. Responses of wildlife are dynamic and keyed into features of the forest vegetation and environment. In assessing the balance of economic harvesting and wildlife conservation, three basic questions must be considered:

1. How is a defined community affected by a particular logging operation and how well does it recover over time? As a corollary to this, what is the effect of repeated logging?
2. Is success in community recovery dependent upon the accessibility of unlogged forest as a source of colonists for those species unable to persist in recently logged forest? If so, what are the requirements for unlogged refuge areas; should they be demarcated on logging concessions and how large should they be?
3. How can existing data on these responses of communities to existing timber production systems be used to help modify these systems, making them both economically and ecologically sustainable?

These questions may all be addressed in the light of current data. Opinions can be formulated, based on the available information, concerning both failings of current production systems and their less often reported favourable aspects. It must be borne in mind, however, that irrefutable proof of the viability or inviability of both current systems and suggested modifications do not yet exist. Experimentation is required to provide objective information.

The timber industry everywhere is coming under tighter controls and foresters are paying more attention to preservational, as well as conservational, aspects of forest management. In pointing out both the deleterious and the favourable aspects of current forest management practices, as far as wildlife populations are concerned, it is to be hoped that forest managers will be moved to address the former.

Before discussing the changes in biological diversity caused by current selective logging and forest management practices, however, certain omissions from the database should be mentioned. First, the numerically most important group, the invertebrates (notably the insects) have not been studied. Almost all available information concerns vertebrates. Invertebrate groups are far more species-rich than vertebrates and are likely to be severely affected by current logging practices. Second, no available data cover a complete logging cycle. It is possible to show evidence in support of various theories as to what will happen in a rain forest subjected to cyclic logging, but there is no objective information. There is no evidence to suggest that wildlife cannot persist over multiple logging cycles, although results suggest that the survival of all species will be unlikely. On the other hand, there is no evidence to prove that most commercial logging systems so far developed are sustainable indefinitely, and wildlife can be expected to persist only under sustainable harvesting. Conclusions in this debate must be contingent upon further research, and the form that this further work might take will be discussed.

Animal population dynamics

Comparative data
Discussion of the responses of animal communities and of individual species to logging operations and any subsequent forest management exercises has generally been based on comparative results from areas of unlogged forest and other areas logged at different times in the past. There is a considerable volume of data of this type for the more conspicuous forest vertebrates (e.g. primates: Johns and Skorupa, 1987).

The greater the species-richness of a species group, the more sensitive it is likely to be to habitat change and thus the more useful as an indicator of changes in ecosystem processes. Studies of invertebrates are likely to demonstrate most conclusively whether or not managed forests are regaining a community equivalent to that of unlogged forest. Data available suggest that considerable changes occur in the community as an effect of forest harvesting. Usually these take the form of drops in the number of species and frequently also in the numerical dominance of a very few species. Short-term moth-trapping programmes carried out in Sabahan dipterocarp forest have demonstrated a considerable difference in species composition between unlogged and logged forest, and have suggested that this is extreme among small less mobile species which cannot move regularly between different habitat types (Henwood, 1986; A. Kirk-Spriggs, pers. comm.). Among termites, preliminary studies indicate that environmental changes cause a considerable drop in species richness, favouring mound-building species over those constructing simple nests in the soil. The result is dominance by a very few species in logged forest (Table 2.3).

The appearance of overdominant species is a common phenomenon in

Table 2.3 Number of termite species recorded in one-day samples in unlogged and logged dipterocarp forest, Sarawak

Termite group	No. species	
	Unlogged	Selectively logged
Rhinotermitidae		
Heterotermitinae	1	0
Coptotermitinae	1	0
Rhinotermitinae	4	0
Termitidae		
Termitinae	11	4
Macrotermitinae	4	3
Nasutitermitinae	4	3
Total species found	25	10

Source: Collins (1980).

logged forests. For example, in a central American bat fauna, this was especially marked with two species of one genus making up three-quarters of a total netted sample (Table 2.4). Similarly, in southern India, a few species of birds tended to dominate a number of disturbed forest avifaunas (Beehler *et al.*, 1987), with the effect of changing estimates of species diversity and equitability.

The use of simple diversity indices, and their interpretation, should be cautioned at this point. There is a strong tendency in the literature to use such simple indices to examine differences between habitats, notably when comparing bird populations. Bird species' diversity correlates strongly with vegetation height diversity (MacArthur *et al.*, 1962); structurally complex vegetation types buffer the effects of seasonality such that resources become predictable (Karr, 1976). In the case of disturbed habitats, such simple indices merely express obvious changes in the vegetation resulting from disturbance. A simple index overlooks the fact that different subsets of the fauna react to disturbance in different ways. As discussed by Connell (1978), disturbed habitats may show a higher species diversity. This does not mean that, in addition to new species, they will continue to support all species of undisturbed forest.

A technically robust way of exploring reasons for difference in species responses to logging has been explored by Skorupa (1986). Skorupa, working with a community of primates, determined by an analysis of covariance that the best predictors of the abundance of species were unique to that species: some were linked to the abundance of particular sets of food trees and others to composite measures of forest structure. He

Table 2.4. Catches of frugivorous or largely frugivorous bats, Chiroptera, in undisturbed and disturbed forests in Panama

Species	No. of individuals netted	
	Bohio Peninsula (undisturbed)	Buena Vista (disturbed)
Phyllostomus discolor	0	2
Carollia castanea	33	23
C. perspicillata	9	73
Uroderma bilobatum	2	12
Vampyrops helleri	1	1
Vampyressa pusilla	9	0
V. major	4	0
Chiroderma villosum	1	0
Artibeus jamaicensis⎫ *A. lituratus*⎭	37	11
Total	96	122

Source: R.J. Pine and D.E. Wilson (unpublished data); reproduced with permission.

hypothesized that densities of primate species closely evolved with primary forest should show a strong positive covariation between unlogged and logged forest plots, and thus that analysis by way of an interspecific correlation matrix should reveal which species, if any, are limited to primary forest communities. Data from primate communities in two geographical areas can be represented pictorially in dendrograms (Figs. 2.4 and 2.5). It can be seen that only in the African community can certain species not be expected to persist alongside current forest management practices. These conclusions have been tested for some of the species concerned (Skorupa, 1986; Johns, 1986b) and found to agree with actual responses to logging.

In many cases it may thus be possible to make predictions based on comparative data, as to the effects of projected forest management practices. Responses to the changed conditions of logged forest can be unexpected, however. For example, hummingbird communities in neotropical rain forest are generally coadapted with communities of flowering plants, with the different species of hummingbird diverging in bill shape to exploit flower species of particular corolla design. In regenerating logged forest, much of the available nectar is present in colonizing shrubs and climbers with flowers of generalized corolla shape (Feinsinger, 1976), with

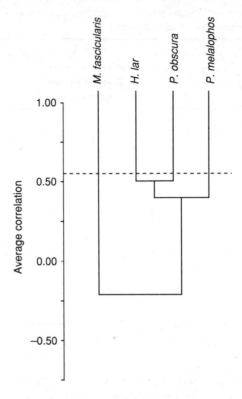

Figure 2.4 Dendrogrammatic representation of primate density interspecific correlation matrix for 11 study plots in Peninsular Malaysia (data from Marsh and Wilson, 1981). The dashed line indicates the 5% probability level for *r* (product-moment correlation coefficient). Species closely coevolved with mature forest should branch off at or above the dashed line. Species that benefit from logging should branch off in negative correlation space. The species illustrated here are from the genera *Hylobates*, *Macaca*, and *Presbytis*. From Skorupa (1986).

the result that social organization of the hummingbird community changes to one of aggressive defence of resources (interference competition). Hummingbirds in logged forest are therefore generally transient; species richness may drop considerably in small isolated forest fragments (Willis, 1979) which might not be expected from examinations of ecological attributes of the species concerned.

Long-term studies
A failing of comparative data collected from different sites is that there can often be wide variation in species abundance, or even in species distribution, over even quite small distances. Analysis of comparative information

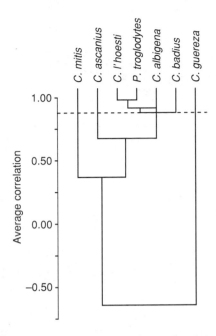

Figure 2.5. Dendrogrammatic representation of primate density interspecific correlation matrix for five study plots in Kibale Forest, Uganda. The dashed line indicates the 5% probability level for r (product-moment correlation coefficient). Species closely coevolved with mature forest should branch off at or above the dashed line. Species that benefit from logging should branch off in negative correlation space. From Skorupa (1986).

assumes an equal basal population, which is generally erroneous. For example, in a Peninsular Malaysian site it has been shown that intersite variation in primate densities is greater than intrasite variation over time, even when the time period spans a logging event (Johns, 1989). This is likely to be especially true of species-rich communities, in which presence–absence or abundance statistics derived from different forest areas are likely to be of limited value. Only by continuing studies in a single forest area throughout a logging and management cycle can entirely objective information be gained concerning their effects. Some information has been collected on the regeneration of Australian tropical forest following intensive logging and burning (Smith, 1985), but only two studies, one in Africa and one in Peninsular Malaysia, have been directed towards monitoring the recovery of the forest fauna following commercial logging operations.

In Peninsular Malaysia, primate densities estimated on two occasions have shown few consistent trends, other than an increase in the abundance of one species which is a specialist of high productivity secondary habitats

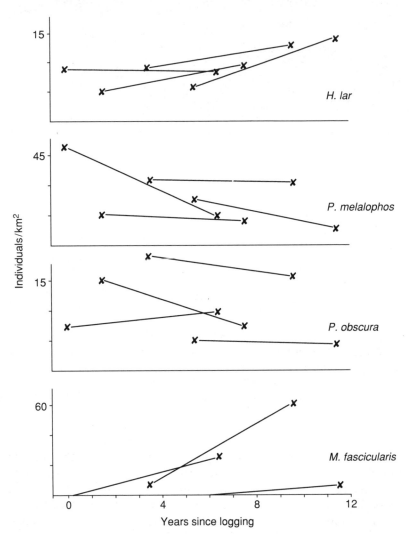

Figure 2.6. Estimated densities of primate species in logged-over forests at Tekam F.R., Peninsular Malaysia. From Johns (1989).

(Fig. 2.6). Primates often take a long time to respond to environmental changes, however. From an examination of infant–female ratios (Table 2.5) it appears that in the long term some species may be better able to survive than others. A lack of long-term changes in primate densities has also been recorded in the Ugandan study (Howard, 1986).

The avifauna of a Peninsular Malaysian site has been found to regain all but five of the 193 species found in unlogged forest by 10 years after the

Table 2.5 Estimated infant–female ratios among primates following selective logging at Tekam F.R., Peninsular Malaysia

| Species | *Infant–female ratios* | | | |
	Unlogged forest	*During logging*	*6 months after logging*	*6 years after logging*
Hylobates lar	0.50	0	0.50	0.33
Presbytis melalophos	0.41	0	0.13	0.14
P. obscura	0.31	0	0.09	0.20
Macaca fascicularis	0.28	0.18	0.18	0.25

Source: Johns (1989).

logging event (Johns, 1989). In Queensland no bird species at all were lost during a logging operation (Crome, 1991). Statistics of this type are commonly published, and can be misleading. They may be taken to support the likelihood of the community persisting alongside the management operation indefinitely. If results of long-term monitoring of the bird community are analysed, however, a different picture emerges. An Index of Overlap (Horn, 1966) suggests that not only do the logged forests possess a different avifaunal community structure, but that it shows no sign of reconverging with that of the unlogged forest over time (Table 2.6). Examined in terms of the abundance of feeding groups, it is also apparent that there are changes in logged forest (Table 2.7) with species of a few groups becoming dominant. The number of species recorded in the sample declines, a reflection of the lesser numbers of rare species and consequently higher equitability typical of disturbed systems. Older logged forests show evidence of the re-establishment of many of the rare species (Johns, 1989), thus equitability would be expected to decrease as the forest regeneration process continues. The structurally complex vegetation in tall rain forest does not allow dominance by a few bird species.

Intensification of logging

Studies in Peninsular Malaysia and elsewhere have suggested that although a loss of 50% of trees may have considerable initial effects on vertebrates, the level of resilience shown by the community was remarkably high. Almost all vertebrate species present in unlogged forest had persisted or had recolonized by 12 years after logging (although the relative abundance of animal species was different in logged forests). There are, of course, some species which are susceptible even to moderate levels of disturbance:

Table 2.6. Similarity of unlogged and logged forest avifaunas as described by Horn's Index of Overlap (Tekam F.R., Peninsular Malaysia). An index of 1.00 indicates an identical avifauna, an index of 0 indicates no overlap

	Area and time elapsed since logging (years)							
Area	C13C (0)	C5A (1–2)	C1A (3–4)	C2 (5–6)	C13C (6–7)	C1A (7–8)	C1A (9–10)	C2 (11–12)
C13C	–							
C5A	0.10	–						
C1A	0.38	0.43	–					
C2	0.04	0.71	0.49	–				
C13C	0.07	0.22	0.33	0.37	–			
C5A	0.09	0.66	0.49	0.78	0.46	–		
C1A	0.36	0.19	0.52	0.29	0.32	0.57	–	
C2	0.09	0.38	0.23	0.46	0.32	0.32	0.29	–

Source: Johns (1989).

an example would be pithecine seed predators in Amazonian forests (Table 2.8). In general, however, the loss of 50% of trees might not disrupt the ecosystem beyond its ability to recover, given sufficient time.

In a few rain forests, logging levels are already in excess of 50% trees removed. It might be supposed that the abundance of many animal species would be correlated with damage levels. In the case of primates there is no significant relationship between the population densities of common primates, or of their total biomass, and the degree of damage within various forests of south-east Sabah (damage being expressed in terms of the basal area of mature forest trees remaining: Fig. 2.7). A lack of correlation is also seen in the density of frugivorous birds and the abundance of food trees: hornbill biomass/km^2 in south-east Sabahan forests is approximately equal over a range of forests from unlogged to heavily logged, despite a loss during logging of up to 73% of the original food trees (Johns, 1989). Small, less mobile species are more likely to suffer population reductions in heavily damaged forests, however. The patchy distribution of food sources may affect ranging patterns, breeding success and even gene flow, unless these species are able to re-occupy the regenerating forests and restore an even dispersion of individuals.

It is perhaps likely that the number of species whose populations are critically reduced by logging is related in some way to the severity of disturbance. It may be predicted that species loss is a logarithmic function: at a high level of timber extraction and associated damage, the loss of a few more trees due to carelessness or bad management practices will cause

Table 2.7. A comparison of feeding guild membership within samples of 800 individual birds observed at one site over time (Tekam F.R., Peninsular Malaysia)

Trophic group	Feeding guild	% sample (n = 800)		
		Before logging	1–6 months after	6–7 years after
Frugivores	Terrestrial	0	0.4	0
	Arboreal	10.6	10.6	14.9
Faunivore/frugivores	Arboreal	11.5	10.3	5.8
Insectivore/frugivores	Terrestrial	0.6	0	0
	Arboreal	15.1	8.8	39.2
Insectivore/nectarivores	Arboreal	3.9	2.0	4.5
Insectivores	Terrestrial	3.6	1.4	0
	Bark-gleaning	5.3	3.4	3.0
	Foliage-gleaning	31.1	34.5	28.5
	Sallying	16.2	24.1	2.0
Carnivores	Raptors	2.0	4.5	2.0
	Piscivores	0.1	0	0
No. species in sample		120	100	86

Source: Johns (1989).

Table 2.8. Relative densities of pithecine primates in unlogged and logged forests of Brazilian Amazonia

Species and area	No. individuals/km²		
	Unlogged	Lightly logged[1]	Heavily logged[2]
Southern bearded saki *Chiroptes satanas*			
Tucuruí	9	–	0
Gorupí	30	15	0.2
White uakari *Cacajao calvus*			
Mamirauá Lake	16	24	< 10

Source: Johns (1986a).
[1] < 15% of trees destroyed.
[2] > 50% of trees destroyed.

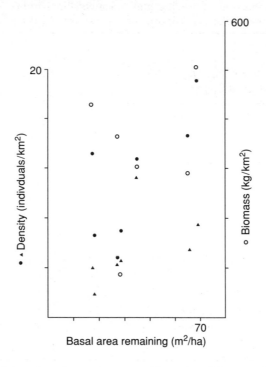

Figure 2.7. Relationships between primate density and biomass and damage levels during logging (Ulu Segama F.R., Sabah). (○ = *Presbytis rubicunda*, △ = *Hylobates muelleri*). From Johns (1989).

disproportionately more losses than would be the case under low logging levels. Following this argument, it may also be predicted that repeated logging, especially for small-sized trees, or re-logging before full regeneration has taken place, is likely to have severe and permanent effects on biological diversity.

Colonization of logged forest

Animal population dynamics in patchy habitats are poorly understood: mechanisms whereby recolonization does or does not occur have been little explored. Of course, suggestions exist as to how to design the habitat disturbance so that it minimizes effects upon biological diversity. Many such suggestions take into account the influence of edge effects in suggesting the sequence of patch-cutting within a concession area (the 'long-rotation island' approach, Harris, 1984; see also Hartshorn, 1990). Unfortunately, biologically attractive management procedures are rarely equally attractive to forest economists (an exception may be the

Queensland model: Nicholson *et al.*, 1988). The majority of managed rain forests are not, and are unlikely to be, harvested in ways favourable to the recolonization of species. The recolonizations of those species intolerant of conditions in recently logged forest is therefore an important subject of study.

Depending on the degree of environmental change caused by logging, a certain number of species are able to persist within the logging area. Only a proportion cannot persist and must therefore recolonize if the system is to be maintained in perpetuity. Failure of recolonization may be due either to a continuing inability of the regenerating habitat to support these species, or to an absence of a source-pool of colonists. Should recolonization fail, it may be taken as evidence of a permanently altered ecosystem and, probably, a forest management system that is not sustainable.

The fundamental question to examine when considering the colonization of logged-over forest by animals is whether, in fact, they ever actually leave and thus have to recolonize. Most mobile species can be expected to adjust activity patterns to at least some extent to help cope with changes in the distribution of food sources, etc. This appears to be the case with many larger animals, such as primates and frugivorous birds. Not all birds can move long distances on a regular basis, however. Some idea of the dispersal ability of birds can be gained from a consideration of the relationship between body weight and wing length (wing-loading). Differences between sedentary terrestrial birds such as ground babblers and pittas, and typical colonists, such as bulbuls, are clear (Fig. 2.8). The former would be expected to be less able to move through logged forest, the more so because they are often physiologically intolerant of sunlit conditions and unwilling to cross open gaps in the vegetation (Johns, 1989).

The proportion of vertebrate species within a community that cannot be expected to survive in forests logged at a moderate intensity may not be high. Many rain forest species, even those of species-rich groups, may show low habitat specificity. This may be determined by such measures as Levins' (1968) Index of Niche Breadth. Except where sample size is very low, few species within a random sample of Amazonian birds showed a low level of tolerance (Table 2.9); most were found across a wide range of the undisturbed and disturbed habitats considered (Johns, 1986a).

The lack of habitat specificity shown by many vertebrate species, and the fact that most intolerant species have very small area requirements (Johns, 1985, 1989) suggests that the provision of quite small unlogged forest refuge areas within logging concessions might be sufficient to maintain populations of most species. Even in heavily logged forests of Sabah with up to 73% of trees lost, most animal species are able to persist either within the logged-over forest itself or within lightly damaged non-commercial stands on slopes and plateaux and in riverine strips left

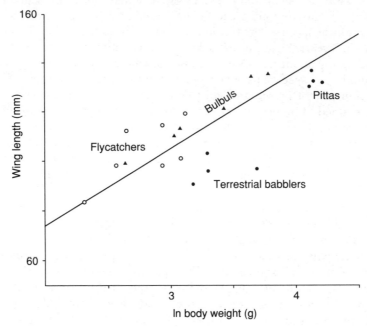

Figure 2.8. Wing-loading of selected species groups of Malaysian understorey birds. The linear regression line is calculated from figures for all species in the community.

unlogged as a requirement of the forest management system.

In another study in Queensland, many bird species were found to make intensive use of streamside vegetation in unlogged forest; the provision of riverine refuge strips in logged forest was shown to be vital to the persistence of many birds, but if even small areas were preserved no bird species were lost following logging (Crome, 1991).

Of course, the extent to which species are able to persist within logged areas may ultimately depend upon the degree of damage over time. Equally, the larger the unlogged refuge area, the more invertebrate and vertebrate species it may support independently of access to surrounding logged forests. Given the economic realities, however, it should be stressed that the provision of even very small unlogged areas is of considerable benefit in maintaining the biological diversity of managed forests.

Measures of sustainability

Long-standing ideas as to what constitutes sustainable logging, such as the simplistic definition as an operation where offtake does not exceed natural

Table 2.9. Inter-habitat tolerances of selected Amazonian bird species (Levins' Index of Niche Breadth)

Species	Feeding guild[a]	Total no. observations	Index of Niche Breadth
Aratinga leucophthalmus	AF	730	0.33
Pionus menstruus	AF	387	0.55
Melanerpes cruentatus	BGI	178	0.34
Tangara mexicana	AIF	171	0.54
Ara macao. chloroptera	AF	155	0.65
Myrmotherula axillaris	FGI	153	0.38
Phaethornis philippi	IN	112	0.52
Crypturellus variegatus	TIF	92	0.39
Penelope jacquacu	TF	84	0.65
Pipra coronata	AIF	65	0.56
Volatinia jacarina	S	50	0.20
Sporophila lineola	S	30	0.20
Cyanerpes caeruleus	AIF	28	0.82
Gymnopithys salvini	TI	27	0.44
Eubucco richardsoni	AF	26	0.61
Leptotila verreauxi	TF	15	0.38
Mitu mitu	TF	8	0.58
Leptotila rufaxilla	TF	2	0.20

[a] Feeding guilds are as follows: TF, terrestrial frugivore; AF, arboreal frugivore; TIF, terrestrial insectivore/frugivore; AIF, arboreal insectivore/frugivore; IN, insectivore/nectarivore; TI, terrestrial insectivore; BGI, bark-gleaning insectivore; FGI, foliage-gleaning insectivore; S, seed-eater. Data from Johns (in press).

increment, are being questioned not only by foresters but by the entire community of people concerned with forest land use. In many areas, careful attention is being paid to recruitment, growth and yield following logging. It is now generally realized that current logging practice almost invariably causes extensive environmental damage. This is, however, at least partly due to flouting of general rules of the management systems employed. Proper adherence to these rules is a prerequisite of sustainable management, but they have so far proven difficult to enforce.

Information on the success of management practices continues to be collected by a number of national forestry departments, but timber logging operations are not being suspended pending the results of such forestry studies. A more rapidly attainable assessment of certain ecological characteristics of the disturbed forest communities may perhaps predict whether disturbance levels preclude the sustainability of harvesting.

Assessments of biological diversity may aid foresters in determining harvesting systems for remaining forests.

Species-abundance distributions

One statistical technique that has been applied is the use of species-area or species-abundance distributions (Stenseth, 1979), which may indicate stability characteristics of an ecosystem. The former is probably more robust, but the latter are easier to obtain and may offer some useful suggestions (Johns, 1989). In brief:

1. A stable community is predicted to show a negative correlation between the number of species present and the variation in their densities. This shows up as a log-normal species-abundance distribution.
2. Environmental disturbance is predicted to amplify variations in the relative densities of species. This results in a log-series species-abundance distribution.

A recently logged forest is expected to show the latter species-abundance distribution and a return to the former is expected to indicate constancy (stability) of environmental conditions, which should allow the regeneration of the original animal community. Data are available for birds from two sites, Peninsular Malaysia (where tree loss was 51%) and Sabah (73% tree loss) (Table 2.10). There is no evidence of a return to a log-normal species-abundance distribution even 11–12 years after logging, but a lessening fit to a log-series curve in Peninsular Malaysia might suggest the

Table 2.10. Species-abundance distributions of Malaysian bird communities in unlogged and logged forests

	Regression coefficient	
Community and forest type	Log-normal	Log-series
Tekam F.R., Peninsular Malaysia		
Unlogged	0.99	0.96
1–2-year-old logged	0.93	0.97
5–6-year-old logged	0.86	0.95
11–12-year-old logged	0.75	0.85
Ulu Segama F.R., Sabah		
Unlogged	0.95	0.88
6-year-old logged	0.89	0.91
12-year-old logged	0.87	0.95

Source: Johns (1989).

beginning of a recovery to a point of stability. There is no sign of this happening in the Sabahan sample.

It is tempting to suggest that re-establishment of a log-normal distribution in logged forest is indicative of successful regeneration and that, in such cases, the community under consideration will accrue successfully to the original community. There is no evidence of this *per se*, although constancy of environmental conditions is one of the prerequisites for this to occur. It should be true, however, that a community in old logged forest that continues to show a log-series species-abundance distribution has not regained a stable state. Re-logging occurring at this stage will have a further degradative effect on the ecosystem.

Individual species populations

In a consideration of results generated from model ecosystems, Pimm (1979) has stressed that large disturbances affecting complex ecosystems are most likely to cause reduction in numbers rather than widespread species deletions. Complex ecosystems will better tolerate the effects of any loss of individual species, except where the species lost is a predator (Pimm, 1984). Since few predators are eliminated by logging operations in rain forest, the absence of an animal species following logging can usually be attributed to environmental factors.

In practice, the precise environmental factors to which species may be responding are difficult to define. The abundance of individual species may be more dependent on such indefinable factors than upon the structure of the community. Only in a few cases may the absence of individual species be correlated with, for example, eutrophication of the water system, or critical levels of soil compaction, or the slowing-down of leaf-litter decomposition. It is also unlikely that groups of species will be accurate indicators of stability characteristics of the ecosystem.

Forest managers should therefore be cautious in examining patterns of individual species abundance. It is tempting to try to define 'indicator species': single animals whose presence, absence or abundance may be used in a quick assessment of the status of a regenerating forest. This is inadvisable because individual species generally do not respond in a consistent manner to broad environmental parameters. There are too many different factors at play at the microhabitat level.

BIODIVERSITY CONSERVATION AND SUSTAINABLE MANAGEMENT

The importance of sustainable management

Forests need to be managed on a sustainable basis for environmental protection and as a source of forest products, including timber. Viable forest management units are necessarily large and will contain a variety of forest wildlife. Sustainable forest management, which should largely mimic natural regeneration processes (Skorupa and Kasenene, 1983; Whitmore, 1990) will be compatible with the maintenance of much biological diversity, although not necessarily with the preservation of all species. In managed forests the proportion of commercial timber species in the stand will often be increased (Wyatt-Smith, 1987b). In South-east Asia, Johns (1989) has demonstrated that as long as the diversity of forest vegetation is maintained, total biological diversity is not greatly affected by a single cycle of logging. If sustainable forest management systems were a genuine priority of the tropical timber industry, then prospects for the long-term conservation of rain forest species would be greatly improved: production forests would then form a complement to totally protected areas in species conservation programmes. It is therefore important that wildlife conservationists promote sustainable forest management systems.

Rethinking current practices

Felling

It is generally recognized that damage levels attached to current logging operations, particularly intensive logging operations, are unnecessarily high. Any forest management techniques subsequently employed are handicapped by the damage already inflicted upon advanced growth and by various environmental problems, such as changes in microclimatic conditions. Experimental operations (such as that of Marn, 1982, in Sarawak, and Malvas, 1987, in Sabah) have shown that the levels of incidental damage, even under intensive harvesting, can be reduced by as much as half.

There is little doubt that the felling operation is the critical part of a forest management procedure. The more time and care that is taken in selecting trees to be felled, marking specimen trees and advanced growth for retention, laying a road network that minimizes damage, controlling directional felling of trees, and so on, the greater the likelihood of success of the management operation as a whole. The net growth rate of the residual stand is variable, but it is more likely that an unfavourable rate of basal area growth will occur if felling damage is high (Table 2.11). Low growth rates will be reflected in low levels of accumulation of commercial timber volume.

Furthermore, there is an optimum canopy gap size at which the regeneration of timber trees may be affected. Larger gaps caused by felling too many adjacent trees commonly stimulate the growth of commercially useless pioneer tree species, woody climbers and shrubs, which can be a

Table 2.11. Net growth of the residual stand as a function of logging intensity (Indonesia)

% of trees destroyed during logging	Years post-logging	Net growth rate $(m^2/ha/year)$
4	7	3.9
9	7	5.9
10	7	3.4
15	7	4.9
20	7	0.8
23	7	−1.7
36	2	−2.4
76	2	−16.3

Source: Miller (1981).

bane to further management operations (as with the climber *Mezoneuron* in Sabah: Whitmore, 1984). Gap size can be controlled during the felling operation to match growth characteristics of desired timber species (either advanced growth that can be retained or seedlings that can be planted out). This is the fundamental basis throughout the world of all natural forest silviculture. In Uganda, for example, detailed information exists concerning the success of commercial tree seedlings in differently sized gaps (Kasenene, 1987) and this information is to be used in developing future felling strategies.

Post-felling

The extent to which it is desirable or feasible to apply post-felling timber stand management to promote the development of a new crop varies according to forest type. It depends especially on the original density and the regrowth characteristics of desired timber trees. In very lightly felled forests, a total absence of silvicultural treatments is, in most cases, desirable. In very heavily logged forests of South-east Asia, it is generally accepted that post-felling treatment is not feasible: economic analyses have shown such operations not to be cost-effective (e.g. Chai and Udarbe, 1977). Post-felling poisoning of non-commercial trees is still promulgated in the CELOS system of Surinam. This is necessary to improve the low commercial volume and release advanced growth of commercial species. Seedlings of these are not light demanding and do not colonize open gaps (De Graaf, 1986). This form of management is intensive and requires a high level of training of forestry staff, but would seem to be sustainable.

In most tropical forests, however, the most effective form of forest management is undoubtedly protection and encouragement of advanced growth in optimally sized gaps created during logging, with planting of gaps where no advanced growth exists. Native species should, of course, be used for this enrichment planting wherever possible. Fast-growing exotics, such as *Acacia mangium* and *Eucalyptus* are often planted in South-east Asia to bind heavily damaged soils in danger of erosion. This would not be necessary if damage to these soils had been minimized by proper road construction and timber extraction practice.

There have been suggestions that the diversity of wildlife may be encouraged by the retention of non-commercial or commercial trees of particular importance to wildlife as food sources. Fig trees (*Ficus*) are often put forward as a case in point (e.g. Leighton and Leighton, 1983). Studies of wildlife in logged forest suggest that enough such trees generally survive to maintain populations of wildlife, with a few exceptions (Johns, 1989). The avoidance of such trees during felling would certainly be of benefit to vulnerable species, however. The development of sustainable operations should include identification of 'key' food trees, and mature species of these should be marked for retention in prelogging enumerations.

Finally, studies in Sabah have demonstrated the importance of the retention of unlogged forest patches within logging areas as 'refuge' areas for the vertebrate species not able to persist in recently logged areas (Johns, 1989). These areas need not be large. Strips of forest left along watercourses (designed to protect water quality) and non-commercial stands, together occupying around 5% of a concession area in Sabah were found to contain populations of most species regarded as intolerant of conditions in recently felled forest (Johns, 1989). The retention of even quite small areas is of extreme importance, but, of course, the larger the refuge areas the greater the probability of successful recolonization and the lower the probability of loss of genetic variability. The allocation of at least 20% of production forest areas as undisturbed nature reserve is planned in Uganda (Tabor *et al.*, 1990), an example of the type of commitment that should be sought.

Further research

Detailed information is now available from several geographical regions concerning the influence of logging on the behaviour and distribution of mammal and bird species. As a result, it is possible to predict the effects upon vertebrate communities of any defined logging operation (given details of timber volumes and species extracted, and levels of damage incurred). The species that will or will not persist under such defined felling and management regimes can be determined in advance of the onset of logging. Far less information is available concerning invertebrate and plant communities, and fewer predictions can be made.

Since it is possible, at least for vertebrates, to predict with some degree of accuracy which species are vulnerable to timber harvesting and management operations, and why, it is also possible to devise management systems that contain elements designed to help such species persist. Taking this a stage further, it may also be possible to devise an economically feasible management operation, in any rain forest type, that optimizes both timber production and the conservation of biological diversity. A system of this type, and only a system of this type, will be ecologically sustainable.

Information on long-term responses of wildlife to logging systems of known form is continuing to be acquired from a number of regions, notably Peninsular and eastern Malaysia, Uganda and Queensland. Less long-term information is available from the Neotropics, an omission that needs to be addressed. Such information is needed to refine the accuracy of predictions concerning the survival and persistence of vertebrate wildlife, and to address species groups about which less is known. Perhaps the most important research yet attempted is about to begin in Uganda, where experimental commercial forestry operations have been designed by a team of wildlife and forest ecologists. The projected forest management operations represent a re-evaluation of standard goals, emphasizing the need to maintain the forest environment and biological diversity as the critical component of a wood production system (both timber and fuelwood). Intensive research on ecosystem processes under different levels of offtake, control of gap size, replanting, regeneration, and so on, coupled with economic analyses, are intended to provide the first objective information concerning the mechanisms whereby a sustainable logging system can be developed.

SUMMARY

Timber harvesting is, and is likely to continue to be, a major use of tropical forests. It will affect most of the forests lying outside totally protected national parks and reserves in the next few decades. Logging is widely seen as constituting a major threat to the biological diversity of tropical forests. The present study shows that many species of vertebrates are able to persist in logged-over forests. However, in some cases many years elapse before certain species recolonize forest even when a source of colonists exists in nearby unlogged forest. Recent research has shown that a high complement of mature forest species can be maintained in production forests if ecologically sustainable modes of harvesting and silviculture are applied.

REFERENCES

Abdulhadi, R., Kartawinata, K. and Sukardjo, S. (1981) Effects of mechanised logging in the lowland dipterocarp forest at Lempake, East Kalimantan. *Malaysian Forester*, **44**, 407–18.

Asabere, P. (1987) Attempts at sustained yield management in the tropical high forests of Ghana, in *Natural Management of Tropical Moist Forests* (eds F. Mergen and J.R. Vincent), Yale University Press, New Haven, pp. 47–70.

Beehler, B.M., Krishna Raju, K.S.R. and Ali, S. (1987) Avian use of man-disturbed forest habitats in the Eastern Ghats, India. *Ibis*, **129**, 197–211.

Bullock, S.H. (1980). Impacts of logging in littoral Cameroon. *Commonwealth Forestry Review*, **59**, 208–9.

Burgess, P.F. (1971). Effect of logging on hill dipterocarp forest. *Malayan Nature Journal*, **24**, 231–7.

Chai, D.N.P. and Udarbe, M.P. (1977) The effectiveness of current silvicultural practice in Sabah. *Malaysian Forester*, **40**, 27–35.

Chivers, D.J. (1972) The siamang and the gibbon in the Malay Peninsula, in *Gibbon and Siamang*, Vol. 1 (ed. D.M. Rumbaugh), Karger, Basel, Switzerland, pp. 102–36.

Collins, N.M. (1980) The effect of logging on termite (Isoptera) diversity and decomposition processes in lowland dipterocarp forests, in *Tropical Ecology and Development*, Vol.1 (ed. J.I. Furtado) International Society of Tropical Ecology, Kuala Lumpur.

Connell, J.H. (1978) Diversity in tropical rain forests and coral reefs: high diversity of trees and corals is maintained only in a non-equilibrium state. *Science*, **199**, 1302–10.

Crome, F.H.J. (1991) Wildlife conservation and rain forest management – examples from north east Queensland, in *Rain Forest Regeneration and Management* (eds A. Gomez-Pompa, T.C. Whitmore and M. Hadley), Parthenon Press, Carnforth and Paris, pp. 407–16.

Dawkins, H.C. (1959) The volume increment of natural tropical high forest and limitations on its improvement. *Empire Forestry Review*, **38**, 175–80.

De Graaf, N.R. (1986) *A Silvicultural System for Natural Regeneration of Tropical Rain Forest in Suriname*. Agricultural University, Wageningen.

Erwin, T.L. (1988) The tropical forest canopy: the heart of biotic diversity, in *Biodiversity* (eds E.O. Wilson and F.M. Peter), National Academy Press, Washington, DC. USA, pp. 105–9.

Feinsinger, P. (1976) Organization of a tropical guild of nectarivorous birds. *Ecological Monographs*, **46**, 257–91.

Freezailah Che Yeom (1984) Lesser-known tropical wood species: how bright is their future? *Unasylva*, **36** (145), 3–16.

Hamzah, Z. (1978) Some observations on the effects of mechanical logging on regeneration, soil and hydrological conditions in East Kalimantan. *Biotrop Special Publication*, **3**, 73–8.

Harris, L.D. (1984) *The fragmented forest*. University of Chicago Press, Illinois, USA.

Hartshorn, G.S. (1990) Natural forest management by the Yanesha Forestry

Cooperative in Peruvian Amazonia, in *Alternatives to Deforestation* (ed. A.B. Anderson), Colombia University Press, New York, USA.

Henwood, A. (1986) Moth trapping in the rain forest of Borneo. Unpublished report to Danum Valley Management Committee, Kota Kinabalu, Sabah, Malaysia.

Horn, H.S. (1966) Measurement of overlap in comparative ecological studies. *American Naturalist*, 100, 419–24.

Howard, P. (1986) Conservation of tropical forest wildlife in western Uganda. Unpublished annual report to World Wildlife Fund International, Gland, Switzerland.

IUCN (1980) *World Conservation Strategy*. IUCN, UNEP, WWF, Morges, Switzerland.

IUCN (1987) *Burma, Forest Conservation Issues*. IUCN, Gland, Switzerland.

Johns, A.D. (1985) Selective logging and wildlife conservation in tropical rain forest: problems and recommendations. *Biological Conservation*, 31, 355–75.

Johns, A.D. (1986a) Effects of habitat disturbance on rain forest wildlife in Brazilian Amazonia. Unpublished report to World Wildlife Fund US, Washington DC, USA.

Johns, A.D. (1988a) Economic development and wildlife conservation in Brazilian Amazonia. *Ambio*, 17, 302–6.

Johns, A.D. (1988b) Effects of 'selective' timber extraction on rain forest structure and composition and some consequences for frugivores and folivores. *Biotropica*, 20, 31–7.

Johns, A.D. (1989) Timber, the environment and wildlife in Malaysian rain forests. Final report to Institute of South-east Asian Biology, University of Aberdeen, Aberdeen, Scotland, UK.

Johns, A.D. (in press) Responses of Amazonian rain forest birds to habitat modification. *Journal of Tropical Ecology*.

Johns, A.D. and Skorupa, J.P. (1987) Responses of rain forest primates to habitat disturbance: a review. *International Journal of Primatology*, 8, 157–91.

Jonkers, W.B.J. (1987) *Vegetation Structure, Logging Damage and Silviculture in a Tropical Rain Forest in Suriname*. Agricultural University, Wageningen.

Karr, J.R. (1976) Seasonality, resource availability and community diversity on tropical bird communities. *American Naturalist*, 110, 973–94.

Karr, J.R. and Freemark, K.E. (1983) Habitat selection and environmental gradients: dynamics in the stable tropics. *Ecology*, 64, 1481–94.

Kartawinata, K. (1978) Biological changes after logging in lowland dipterocarp forest. *Biotrop Special Publication*, 3, 27–34.

Kasenene, J.M. (1987) The influence of mechanized selective logging, felling intensity and gap-size on the regeneration of a tropical moist forest in the Kibale Forest Reserve, Uganda. Unpublished PhD dissertation, Michigan State University, East Lansing, USA.

Leighton, M. and Leighton, D.R. (1983) Vertebrate responses to fruiting seasonality within a Bornean rain forest, in *Tropical Rain Forest Ecology and Management*, (eds S.L. Sutton, T.C. Whitmore, and A.C. Chadwick), Blackwell, Oxford.

Leslie, A.J. (1987) A second look at the economics of natural management system in tropical mixed forests. *Unasylva*, **39**, (155), 46–58.

Levins, R. (1968) *Evolution in changing environments*. Princeton University Press, Princeton.

MacArthur, R.H. (1957) On the relative abundance of bird species. *Proceedings of the National Academy of Sciences USA*, **43**, 293–95.

MacArthur, R.H., MacArthur, J.W. and Preer, J. (1962) On bird species diversity. II. Predictions of bird census from habitat measurements. *American Naturalist*, **96**, 167–74.

Malvas, J.D. Jr (1987) Development of forest sector planning, Malaysia: A report on the logging demonstration cum training coupe. UNDP/FAO Field Document MAL/85/004, no.7.

Marn, H.M. (1982) The planning and design of the forest harvesting and log transport operation in the mixed dipterocarp forest of Sarawak. UNDP/FAO Field Document MAL/76/008, no.17.

Marn, H.M. and Jonkers, W. (1981) Logging Damage in Tropical High Forest. Unpublished report, Forest Department, Sarawak.

Marsh, C.W. (1980) Primates and economic development on the Tana River, Kenya: the monkey in the works, in *Tropical Ecology and Development*, (ed. J.I. Furtado) vol.1, International Society of Tropical Ecology, Kuala Lumpur, pp. 373–6.

Marsh, C.W. and Wilson, W.L. (1981) *A survey of primates in peninsular Malaysian forests*. University Kebangsaan Malaysia, Kuala Lumpur.

Miller, T.B. (1981) Growth and yields of logged-over mixed dipterocarp forest in East Kalimantan. *Malayan Forester*, **44**, 419–24.

Nicholson, D.I. (1958) An analysis of logging damage in tropical rain forest in North Borneo. *Malayan Forester*, **21**, 235–45.

Nicholson, D.I., Henry, N.B. and Rudder, J. (1988) Stand changes in north Queensland rain forests. *Proceedings Ecological Society of Australia*, **15**, 61–80.

Pimm, S.L. (1979) Complexity and stability: another look at MacArthur's original hypothesis. *Oikos*, **33**, 76–86.

Pimm, S.L. (1984) The complexity and stability of ecosystems. *Nature*, **307**, 321–6.

Plumptre, R.A. and Earl, D.E. (1984) Integrating small industries with management of tropical forest for improved utilization and higher future productivity. Paper presented to IUFRO Division P5.01 Meeting on Properties and Utilization of Tropical Timbers, Manaus, Brazil.

Poore, D. (1989) *No timber without trees: sustainability in the tropical forest*. Earthscan, London, UK.

Redhead, J.F. (1960) An analysis of logging damage in lowland rain forest. *Nigerian Forestry Information Bulletin* (New series), **10**, 5–16.

Sayer, J.A. and Whitmore, T.C. (1990) Tropical moist forests: destruction and species extinction. *Biological Conservation*, **55**, 199–214.

Skorupa, J.P. (1986) Responses of rainforest primates to selective logging in Kibale Forest, Uganda: a summary report, in *Primates: the Road to Self-sustaining*

Populations (ed. K. Benirschke). Springer Verlag, New York, pp. 57–70.

Skorupa, J.P. and Kasenene, J.M. (1983) Tropical forest management: can rates of natural treefalls help guide us? *Oryx*, **18**, 96–101.

Smith, P. (1985) Effects of intensive logging on birds in eucalypt forest near Bega, New South Wales. *The Emu*, **85**, 15–21.

Stenseth, N.C. (1979). Where have all the species gone? On the nature of extinction and the Red Queen hypothesis. *Oikos*, **33**, 196–227.

Struhsaker, T.T. (1987) Forestry issues and conservation in Uganda. *Biological Conservation*, **39**, 209–34.

Tabor, G.M., Johns, A.D. and Kasenene, J.M. (1990) Deciding the future of Uganda's tropical forests. *Oryx*, **24**, 208–14.

Tang, H.T. (1987) Problems and strategies for regenerating dipterocarp forests in Malaysia, in *Natural Management of Tropical Moist Forests* (eds F. Mergen and J.R. Vincent), Yale University Press, New Haven, pp. 23–46.

Terborgh, J. (1983) *Five New World Primates*. Princeton University Press, Princeton.

Uhl, C. and Viera, I.C.G. (1989) Ecological impacts of selective logging in the Brazilian Amazon: a case study from the Paragominas region of the state of Pará. *Biotropica*, **21**, 98–106.

Whitmore, T.C. (1984) *Tropical Rain Forests of the Far East*, 2nd edn, Clarendon Press, Oxford.

Whitmore, T.C. (1990) *An Introduction to Tropical Rain Forests*. Clarendon Press, Oxford.

Whitmore, T.C. and Silva, J.N.M. (1990) Brazil rain forest timbers are mostly very dense. *Commonwealth Forestry Review*, **69**, 87–90.

Whitten, A.J., Damanik, S.J., Anwar, J. and Hisyam, N. (1984) *The Ecology of Sumatra*. Gadjah Mada University Press, Yogyakarta.

Willis, E.O. (1979) The composition of avian communities in remanescent woodlots in southern Brazil. *Papeis Avulsos. Zool., Mus. Sao Paulo*, **33**, 1–25.

Wyatt–Smith, J. (1963) Manual of Malayan silviculture for inland forests (2 volumes). *Malayan Forest Records*, **23**.

Wyatt-Smith, J. (1987a) The management of tropical moist forest for the sustained production of timber: some issues. *IUCN/IIED Tropical Forest Policy Paper*, **4**.

Wyatt-Smith, J. (1987b) Problems and prospects for natural management of tropical moist forests, in *Natural Management of Tropical Moist Forests* (eds F. Mergen and J.R. Vincent), Yale University Press, New Haven, pp. 5–22.

— 3

How many species will there be?

W.V. REID

INTRODUCTION

There is little doubt that the rate of species extinction has grown during the course of this century, and a consensus exists among scientists that a significant loss of the world's species will occur in coming decades if present trends of tropical deforestation continue. How large is the loss of species likely to be? Although the loss of species may rank among the most significant environmental problems of our time, relatively few attempts have been made to rigorously assess its likely magnitude.

The scant literature on extinction rates is due in part to data limitations and thus a recognition of the imprecision of any estimate. Moreover, it can be argued that the exact rate of extinction is not terribly important given that current extinction rates greatly exceed background rates. For instance, 60 birds and mammals are known to have become extinct between 1900 and 1950 (Fig. 3.1), whereas the background extinction rate for these two

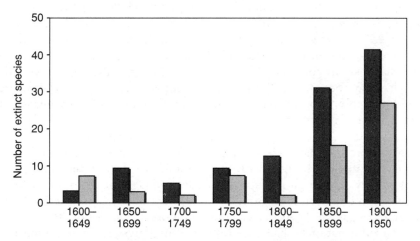

Figure 3.1. Trend in extinctions of bird and mammal species. ■, birds; ▨, mammals. From Reid and Miller (1989).

Table 3.1. Recorded extinctions, 1600 to present

Taxa	Approximate number of species	Total extinctions since 1600
Mammals	4 000	83
Birds	9 000	113
Reptiles	6 300	21
Amphibians	4 200	2
Fish[a]	19 100	23
Invertebrates[a]	1 000 000 +	98
Vascular plants	250 000	384[b]
Total		724

[a] Extinction totals primarily representative of North America and Hawaii (Ono *et al.*, 1983).
[b] Vascular taxa (includes species, subspecies, and varieties).
  Source: Reid and Miller (1989).

groups is only one extinction every 100 to 1000 years based on the average lifespan of a species of 1 to 10 million years (Raup, 1978). Overall, some 724 species are known to have been lost since 1600, and these recorded extinctions are no doubt only a fraction of the total (Table 3.1).

Despite the merits of this argument, the fact remains that most people and policy-makers are unlikely to assess the urgency of today's extinction crisis – and determine the priority given to the issue – by comparing it to background rates of extinction. Instead, priorities for action will be set on the basis of the absolute numbers of species likely to be lost over the coming decades, or on the basis of threats to certain 'flagship' species. A loss of 25% of the world's species over one human lifetime would be staggering by any measure and would readily mobilize political action to slow that loss. In contrast, a loss of 1% of the world's species over the same interval, though equally staggering from an evolutionary perspective, would probably result in the problem being placed on the back-burner of needed social and political change.

While better knowledge of extinction rates can clearly improve the design of public policies, it is equally apparent that estimates of global extinction rates are fraught with imprecision. We do not yet know how many species exist, even to within an order of magnitude (Wilson, 1988); the best data on rates of tropical deforestation are bounded by large uncertainties (WRI, 1990; Sayer and Whitmore, 1990; FAO, 1990); and essential information is lacking on endemism, forest fragmentation, and the potential for species persistence in disturbed habitats. Moreover, ecologists can never even know if extinction rate estimates are borne out. It will not be possible to determine at a future date the number of species that went extinct between now and that future date without knowledge of how many species exist today.

The serious problems associated with extinction rate estimation may explain why the field has advanced relatively slowly since the early calculations of Myers (1979) and Lovejoy (1980), but given the importance of the issue it is surprising that ecologists have not mounted a more substantial effort to estimate rates of extinction and thereby help clarify the actions needed in response. For example, consider the comparable problems and uncertainties associated with predicting rates and impacts of global warming. Policy-makers realize that considerable uncertainty surrounds predictions of climate change, yet they are willing to modify policies based on the consensus views of the scientific community. Although the uncertainty is great, decision-makers accept that the best estimates of reputable scientists are a much better basis for policy formulation than no estimate at all. Accordingly, for the past three years the United Nations Environment Programme and the World Meteorological Organization have facilitated the development of a world-wide scientific consensus on climate change impacts and response options through the IPCC (Intergovernmental Panel on Climate Change).

The issue of climate change contrasts with that of species extinction. Although many ecologists agree that the rate of loss of species in tropical forest is high (but see Simon, 1986; Mares, 1986; Lugo, 1988), the literature addressing this phenomenon is relatively small (Table 3.2). Increased research and debate on this topic would have a considerable pay-off. Efforts to clarify the magnitude of the extinction crisis and the steps that can be taken to defuse the crisis could considerably expand the financial and political support for actions to confront what is indisputably the most serious issue that the field of ecology faces, and arguably the most serious issue faced by humankind today.

ESTIMATING EXTINCTION RATES DUE TO TROPICAL DEFORESTATION

The best tool available to estimate species extinction rates is the use of species–area curves. These curves represent the relationship between the number of species in a region and its area. Based on these curves, it is possible to predict the proportion of species that will become extinct in a region based upon the amount of habitat that is lost. This approach has formed the basis for almost all current estimates of species extinction rates. This technique can either be applied to specific localities in which deforestation rates and species numbers are high (Raven, 1987, 1988a, b;

Table 3.2. Estimated rates of extinction

Estimate	% Global loss per decade[a]	Method of estimation	Reference
One million species between 1975 and 2000	4	Extrapolation of past exponentially increasing trend	Myers (1979)
15–20% of species between 1980 and 2000	8–11	Estimated species–area curve; forest loss based on Global 2000 projections	Lovejoy (1980)
12% of plant species in neo-tropics, 15% of bird species in Amazon basin[c]	–	Species–area curve $(z = 0.25)$[b]	Simberloff (1986)
2000 plant species per year in tropics and subtropics	8	Loss of half the species in area likely to be de-forested by 2015	Raven (1987)
25% of species between 1985 and 2015	9	As above	Raven (1988a, b)
At least 7% of plant species	7[de]	Half of species lost over next decade in 10 'hot spots' covering 3.5% of forest area	Myers (1988)
0.2–0.3% per year	2–3[d]	Half of rain forest species assumed lost in tropical rain forests to be local endemics and becoming extinct with forest loss	Wilson (1988, 1989)
2–13% loss between 1990 and 2015	1–5[d]	Species–area curve $(0.15 < z < 0.35)$[b]; range includes current rate of forest loss and 50% increase	this study

Myers, 1988) or can be applied on a regional (Simberloff, 1986) or global (Lovejoy, 1980; Wilson, 1988; Reid and Miller, 1989) basis.

The following sections update the analysis of Reid and Miller (1989) based on recent evidence that tropical deforestation is proceeding much more rapidly than previously thought. This is followed by a critical examination of the assumptions behind the use of the species–area technique to clarify the limits of this type of analysis, and a brief exploration of the policy options suggested by the analysis as means for lessening the magnitude of the extinction crisis.

Tropical forest habitat

Three parameters are needed to estimate species loss due to deforestation: current forest area, rate of deforestation, and the slope of the species–area curve. Tropical forests can be readily divided into closed and open forest types, based on the density of trees and the presence or absence of a continuous grass cover below the trees (FAO, 1981). Closed tropical forests support the great majority of tropical species and thus it is most meaningful to examine trends of deforestation and extinction in this one forest type. While the area of climates suitable for closed tropical moist forest is 1600 million hectares (Sommer, 1976), part of this is thought to have historically been covered by other vegetation types due to soil type and other ecological and historical factors. Simberloff (1986) estimated that prior to significant human impacts, 10% was not forested, thus the original extent of the biome was 1440 million hectares (Table 3.3).

The most recent estimates of closed tropical forest cover indicate that in the 1980s the total area was 1166 million hectares—roughly 80% of its original extent (Table 3.3). Note that 'closed forest' cover includes more vegetation types than 'closed moist forest', thus the figure of 80% loss is conservative. An estimated 15–20% of this forest area was disturbed by logging or related activities (Lanly, 1982; Reid and Miller, 1989). Land with woody vegetation that was in the fallow portion of shifting cultivation cycles is not included in the total of forest cover.

---

[a] Based on total species number of 10 million. Estimates in bold face indicate the actual loss of species over that time period (or shortly thereafter). Estimates in standard type refer to the number of species that will be committed to extinction during that time period as a new equilibrium is attained.
[b] See text for definition of z.
[c] Extinction estimates apply to the number of species committed to extinction by the year 2000 at current rates of forest loss. How long it will take for the new equilibrium to be achieved is not known.
[d] Estimate refers to number of species committed to eventual extinction when species numbers reach equilibrium following forest loss.
[e] This estimate applies only to hot spot regions, thus the global extrapolation is conservative.

Table 3.3. Area of closed tropical forest (thousand hectares)

Region	Area of suitable climate[a]	Estimated original maximum area[b]	Area in 1980s[c]	Average annual deforestation[c]
Africa and Madagascar	362 000	325 800	203 714	1508
Asia and Pacific	435 000	391 500	288 953	3713
Latin America and Caribbean	803 000	722 700	673 715	5278
Total	1 600 000	1 440 000	1 166 382	10 499

[a] From Sommer (1976). This is area of moist tropical forest, thus does not include all tropical closed forest.
[b] Calculated by subtracting 10% from the climatic climax area (after Simberloff, 1986).
[c] See Appendix. This is area of all closed tropical forest.

The term 'deforestation', as used by forestry statisticians, refers to the transformation of forested land to permanently cleared land or to a shifting-cultivation cycle. In a study conducted in the early 1980s, FAO estimated that annual tropical deforestation amounted to 11.4 million hectares per year (Lanly, 1982). Of this, 7.4 million hectares (65%) of closed forest were lost each year. However, three new estimates of tropical forest loss have recently been released that indicate deforestation is proceeding much more rapidly than previously thought. World Resources Institute has estimated that tropical forests are being lost at a rate of 20.4 million hectares annually, an increase of 79% from FAO's estimate of total forest loss (WRI, 1990). Myers (1989), examined rates of loss of closed tropical forest and concluded that 14.2 million hectares were being lost per year in the 1980s – 90% greater than FAO's estimate of closed forest loss (in part because the studies used different definitions of 'deforestation') (Sayer and Whitmore, 1990). Finally, FAO released updated statistics placing tropical closed and open forest loss at 16.8 million hectares per year (FAO, 1990). The most up-to-date estimate of deforestation rates for closed tropical forest is 10.5 million hectares per year (see the Appendix).

Deforestation rates are imprecise to begin with, and are likely to change significantly in coming decades. In some regions, rates may increase because of population growth or increased access to forest resources. Satellite data for the southern portion of the Amazon basin of Brazil, for example, indicate that deforestation rates rose exponentially between 1975 and 1985 (Malingreau and Tucker, 1988). Eventually, however, rates of forest loss will slow when the most accessible land has been cleared. Accordingly, in order to reflect the considerable uncertainty in predicting future rates of forest loss, species extinction rate is modelled here based on three scenarios of closed tropical forest loss: 5 million, 10 million, and 15 million hectares per year. The low scenario is well below any current estimates, but allows examination of the decrease in extinction rates that could be achieved if a concerted global effort were made to slow tropical forest loss. The middle scenario is comparable to the best current estimate of rates of closed forest loss, and the high scenario exceeds the highest current estimate of closed forest loss.

Species–area curve

Species–area curves generally fit closely to equations in the form:

$$S = cA^z \tag{1}$$

where S = number of species, A = area, and c and z are constants. The exponent z determines the slope of the curve and is the critical parameter in estimating extinction rates.

The estimation of extinction rates is sensitive to the form of the species–area curve. Slopes of species–area curves may differ among various regions of tropical forests because of differences in the numbers of habitats or life-zones present in the region. The slopes are also likely to differ among taxa due to differences in the average size of species' ranges. Curves for groups that tend to have small ranges (high local endemism) should have relatively steep slopes. In addition, species–area curves measured for island flora and fauna differ from those measured for subsets of continental habitats, and the choice of which type of curve to use is not straightforward, as will be discussed further below. Because of the uncertainty associated with the choice of the proper slope for the curve, species extinction rates are modelled over a broad range of slopes that have been found in empirical studies ($0.15 < z < 0.35$; Connor and McCoy, 1979).

Analysis

Extinction rates are estimated for each scenario of deforestation by first calculating regional deforestation rates for Africa and Madagascar, Asia and Pacific, and tropical America assuming that under any deforestation scenario the rate of deforestation for each region would be a constant fraction of the global rate. Initial forest cover in 1990 is then calculated from estimated forest cover in the 1980s (Table 3.3), by subtracting five years of forest loss at the appropriate regional deforestation rate. Then, for each region proportionate species loss is calculated using Equation (1) over time periods of 0–50 years and over species–area slopes in the chosen range of $0.15 < z < 0.35$.

The analysis is performed regionally rather than globally because species diversity differs considerably among the three regions. Africa and Madagascar are believed to contain about 23% of the world's tropical plant species, Asia contains 26% and the neotropics contain 51% (Raven, 1987; Reid and Miller, 1989). Thus, a 5% loss of species in Africa would contribute much less to global species extinction rates than a 5% loss in Latin America. For this same reason it could be argued that the analysis should be performed at an even more fine-grained resolution (i.e., at a country level), but the uncertainties in estimates of both forest loss and species richness at this level of resolution outweigh any theoretical gains in precision.

Global estimates of species extinction rates are then obtained from the regional analyses by weighting each regional estimate by the fraction of the tropical flora occurring in that region. This step assumes that regional patterns of diversity in all tropical forest species parallel the pattern found in plants.

Interpretation

The extinction rates calculated with this model do not represent the actual loss of species over the time period indicated. Rather, they estimate the proportion of species that will eventually go extinct when the system reaches equilibrium following the loss of a given amount of forest. This distinction can best be seen by way of example. Barro Colorado Island in Panama was created around 1914 when the Panama Canal was built and surrounding valleys were dammed. Of the approximately 200 land bird species known to have bred on the island, 47 had disappeared by 1981 (Karr, 1982). The isolation of the island from the formerly continuous habitat committed these species to local extinction, but in many cases the species persisted on Barro Colorado for decades after the island was created. Similarly, estimates derived from the use of species–area curves indicate only the portion of species that will go extinct unless their habitat is restored as species numbers move toward a new equilibrium. Particularly vulnerable species may go extinct immediately and other species with short generation times might be lost within a few years, but some long-lived trees may persist for centuries or even millennia.

The extinction rates predicted by the model for the three regions differ considerably due to regional differences in both deforestation rates and the extent of current forest cover (Table 3.4 and Fig. 3.2). For example, at

Table 3.4. Predicted percentage extinction of tropical closed forest species at equilibrium based on three scenarios of tropical deforestation

Year	Region	Percentage decline in species numbers at equilibrium resulting from deforestation between 1990 and year indicated[a]		
		Low scenario (5 million ha/yr)	Mid scenario (10 million ha/yr)	High scenario (15 million ha/yr)
2015	Africa	1 to 3	3 to 6	4 to 9
	Asia	2 to 5	5 to 11	8 to 18
	Latin America	2 to 4	4 to 8	6 to 13
	All Tropics[b]	2 to 4	4 to 8	6 to 14
2040	Africa	3 to 6	6 to 13	10 to 21
	Asia	5 to 11	12 to 26	28 to 53
	Latin America	3 to 8	8 to 18	15 to 32
	All Tropics[b]	4 to 8	9 to 19	17 to 35

[a] Estimates based on species–area model ($0.15 < z < 0.35$). Current forest area and rates of forest loss from the Appendix.
[b] Total for tropics is the weighted regional average based on fraction of plant species in each region.

deforestation rates of 10 million hectares per year, the average extinction rate for Africa is 1–2% of species per decade, whereas the rate is 2–5% per decade in Asia. Globally, the model predicts that in the next 25 years, current rates of forest loss (approximated by the middle scenario) will commit between 4 and 8% of the world's closed tropical forest species to extinction. These estimates would be reduced to as low as 2% under the low scenario of deforestation, but could be as great as 6–14% of species under the high scenario. History has shown an accelerating rate of deforestation in tropical regions, thus the 'high' scenario may be the most probable in coming decades.

These estimates of species extinction in closed tropical forests can be placed in a global perspective by assuming that the bulk of species loss in coming decades will result from tropical deforestation. Since an estimated 50–90% of the world's species occur in closed tropical forest (Myers, 1980; Reid and Miller, 1989), then at current rates of deforestation the world stands to commit 2–7% of species to extinction in the next quarter century and if rates of deforestation accelerate this loss would increase to as much as 13%. With roughly 10 million species on earth, at current rates of forest loss this would amount to between 8000 and 28 000 species per year, or 20–75 species per day.

The magnitude of this estimate, although lower than some other predictions (Table 3.2), seems to defy common sense and experience. But

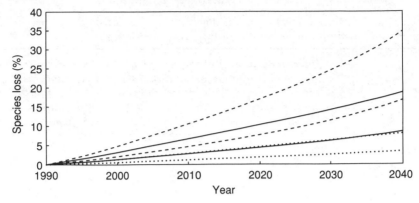

Figure 3.2. Extinction estimates for tropical closed forests. Each deforestation rate scenario was analysed at two values of the slope of the species–area curve ($z = 0.15$, $z = 0.35$). The species loss estimate indicates the fraction of closed tropical forest species that will ultimately go extinct when the system reaches an equilibrium following forest loss of the amount indicated at that time. $\cdots$, 5 million ha/yr; ——, 10 million ha/yr; – – –, 15 million ha/yr.

consider that most of us are barely conscious of the vast majority of species on earth. Large, visible, species like birds, mammals and plants make up less than 5% of the world's species. Considering only these three groups, the results of this study imply that one bird, mammal, or plant will be committed to extinction every 0.5–1.7 days at current rates of forest loss.

ASSUMPTIONS OF THE MODEL

Like any model, the species–area model is a[...]
upon several assumptions. First, the mode[...]
eliminates all species originally present in th[...]
this would overestimate the actual impact o[...]
land under shifting cultivation is considered[...]
dependent species could persist in the wood[...]
Birds and mammals in tropical rain forest ha[...]
habitat that is slightly or moderately distur[...]
agriculture and low levels of selective logging ([...]
2), and deforestation may leave relict fragmer[...]
species continue to survive (Brown and Brows[...]

However, this bias is somewhat offset becau[...]
that species richness in all forested land is at hi[...]
of the forested land included in the calculatio[...]

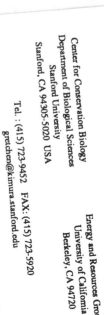

Gretchen C. Daily, Ph. D.

Center for Conservation Biology
Department of Biological Sciences
Stanford University
Stanford, CA 94305-5020 USA

Tel.: (415) 723-9452   FAX: (415) 723-5920
gretchen@kimura.stanford.edu

Energy and Resources Group
University of California
Berkeley, CA 94720

logged, and logging in any significant scale changes habitat structure and species composition. Arguably such logged forest could be considered 'deforested', but more likely the effects on the number of the original species present will be intermediate. For example, one study in Malaysia found that a 25-year-old logged forest contained nearly 75% of the avifauna of a virgin forest (Wong, 1985). In that study the presence of adjacent undisturbed forest probably increased the number of species found in the disturbed site and, moreover, the pattern cannot be readily extrapolated to other groups such as insects. Nevertheless, the study does indicate that more work is needed to clarify the relationship between various types of habitat disturbance and species persistence.

Second, the model assumes that extinction rates are unaffected by forest fragmentation. Obviously, this is unrealistic. In practice, deforestation converts relatively continuous tracts of forest into a fragmented array of smaller patches. Many of the species present in the patches may be lost if their populations are reduced below their minimum viable population size (Shaffer, 1981). But exactly how fragmentation affects species richness within a large region depends upon which areas remain under forest cover. For example, more species may persist if several patches of forest with high species richness or high endemism are selected for protection than if one large tract of similar area but lower richness or endemism is protected (Simberloff and Abele, 1982; Quinn and Harrison, 1988). In general, however, there is every reason to think that by not including the effect of fragmentation, species–area models substantially underestimate the extinctions that will actually occur (Simberloff, Chapter 4). FAO has now begun to include a 'Fragmentation Index' in their reports of forest area which may help ecologists to refine estimates of species loss (FAO, 1990).

Related to this point, the model also assumes that habitat loss occurs randomly among regions with various levels of species richness. If lowland forest sites with high species richness are preferentially deforested, the analysis would underestimate extinction rates. Alternatively, if species–rich sites are preferentially protected, the model would overestimate extinction rates. To circumvent the uncertainties inherent in this assumption, other studies have focused on more detailed assessments of the number of species in regions experiencing high rates of deforestation (Raven, 1988a,b; Myers, 1988).

Finally, the model differs from some applications of the species–area analysis by not assuming either a 'continental' or 'island' slope for the species–area curve. Species–area curves for habitat islands have steeper slopes than curves for subsamples of continuous habitats due to the 'relaxation effect' (Connor and McCoy, 1979). The relaxation effect refers to the gradual loss of species through time after a habitat is fragmented. Many populations are reduced below their minimum viable population size in small habitat fragments. Although these species may persist for varying periods of time after the fragment is created, with time the number of species present decreases until only species with sufficiently large populations remain. Accordingly, although many exceptions exist, the value of the exponent z in the species–area curve for islands generally falls between 0.2 and 0.4 whereas for subsamples of continuous habitat it is less steep, lying between 0.12 and 0.19 (MacArthur and Wilson, 1967; Connor and McCoy, 1979).

Clearly, the goal of this analysis is to examine species loss associated with deforestation, thus fitting the conditions of habitat 'islands' rather than continuous habitats. However, because the model assumes that remaining tropical forest is a continuous, unfragmented tract in each region, the size of each of the three 'islands' greatly exceeds the size of islands where species–area curves have been measured. Even species–area curves that have been measured on continents have involved smaller and generally much more uniform tracts of habitat than would be found in the large areas involved in this study like the Amazon basin.

The choice of the appropriate species–area curve thus becomes quite problematic. Strictly speaking, since the model assumes that the forest is unfragmented and since the area involved is so large, there is little justification for assuming that species number will 'relax' to lower levels with time. Thus a continental species–area curve seems most suited for the analysis. However, because the area involved is so large and heterogeneous, it is not entirely clear if a standard continental curve (measuring primarily beta or 'within-habitat' diversity) is appropriate since considerable gamma or 'between-habitat' diversity would exist. For example, I calculated a species–area curve based on data for plant species richness in 57 tropical countries (data from Davis *et al.*, 1986, supplemented by Gentry, 1986). The exponent z for this curve is 0.30 ($\pm$ 0.066 SE) – falling within the range of 'island' exponents and outside the range of 'continental' exponents.

Added to these issues are the already-mentioned difficulties related to the fact that species–area slopes may differ among regions because of differences in the numbers of habitats present in the region, and slopes are also likely to differ among various taxa due to differences in the level of local endemism. Thus, the best option appears to be to model extinction rates using a broad range of estimates of the slope of the species–area curve.

Because the model assumes that the forest area in each region is continuous, a case could be made that the estimates of species loss apply only to immediate extinctions (due to the complete loss of certain habitats) and do not incorporate relaxation effects. However, in reality, tropical forests are being extensively fragmented and thus it is more realistic to treat the estimates derived as indicating the number of species that will remain when the system ultimately reaches equilibrium.

DISCUSSION

The estimates of extinction rates derived here are comparable to estimates derived in several recent studies (Table 3.2). All these studies have yielded rates in the range of 1–10% extinction (or commitment to extinction at equilibrium) per decade. In some cases these estimates have applied only to plants, but it is reasonable to assume that rates of extinction would be at least as great for all species (since numbers of invertebrate species are even more concentrated in the tropics than plants).

More significantly, although all the estimates in Table 3.2 are based on species–area analysis, two different methods have been used. This study has examined global patterns of forest loss, whereas others such as Raven (1988a, b) and Myers (1988) have focused on specific regions facing serious threats. Accordingly, the general concurrence of results suggests that the conclusions are robust. The fact that the results of this study are somewhat lower than those of Raven (1988a, b) could well be due to the violation of the assumption noted above regarding the pattern of forest loss. If species-rich habitats are being preferentially lost, then the approach used in this study would underestimate global extinction rates. Historical trends provide ample reason to believe that many species-rich regions are under serious threat. For example, over 95% of the species-rich Atlantic coastal forest of Brazil has been lost and less than 9% of the original primary forest of the Philippines remains (Forest Management Bureau, 1988). Myers (1988) has calculated that 10 tropical forest regions, covering only 3.5% of remaining tropical forest contain at least 27% of higher plant species and some 14% of these species are endemic to these regions. Thus, the results of this study may well be conservative estimates of extinction rates. The differences among estimates are due more to methodological than biological factors since key data relating to patterns of fragmentation and endemism are missing.

But especially since they are conservative, the results are cause for alarm. If forest loss continues to accelerate, as many as 35% of tropical forest species may be committed to eventual extinction by the year 2040. Climatic changes associated with the loss of such large areas of forest could exacerbate this potential loss. If instead, a commitment is made to slow deforestation rates to roughly one-half of current levels, the threat to species could drop to well under 10% by the year 2040.

Species loss can be slowed not only by blanket efforts to slow tropical deforestation, but also by protection of key forest habitats with high species-richness and endemism. Tropical forests are being fragmented and lost, but by controlling the pattern and location of fragments and by ensuring that relatively large areas remain in natural and semi-natural habitats, the loss of species could be significantly reduced. Unfortunately, the detailed studies of tropical forest species distributions required to choose such sites are still lacking despite eloquent calls for their implementation made a decade ago (NRC, 1980).

The fact that tropical forests are becoming increasingly fragmented also points to the need for greatly expanded research on the management of small populations. As habitat area is reduced, extinction can be slowed to the extent that managers are able to maintain forest fragments in a supersaturated state; that is, to the extent that species numbers are prevented from 'relaxing' to their natural equilibria.

Species loss can also be slowed by enhancing the conservation of biodiversity in disturbed habitats. As with the need for rapid biological inventories, there is a pressing need for research to identify resource management techniques that will better meet conservation needs under the constraints of timber harvest or rural agriculture. Clearly, regional land-use planning must increasingly incorporate biodiversity conservation as a major planning goal.

A substantial loss of species is already underway, and we are certain to experience species extinctions for decades because of the habitat loss and forest fragmentation that has already occurred. But while the stage is already set for future losses, the magnitude of the extinction can still be influenced by national and international policies and programmes. Many species already 'condemned' to extinction could in fact be saved through restoration of habitat or through new management techniques. But more importantly, a global effort to slow tropical forest loss, rationalize forest use, and maintain key forest habitats could dramatically slow the rate of extinction. Based on the model described above, the 40–90% increase in rates of closed forest loss in the 1980s, if maintained until 2040, will increase the number of tropical species condemned to extinction by between 60 and 160%. These trends must be reversed.

SUMMARY

Recent evidence that tropical deforestation has accelerated in the 1980s has profound implications for the persistence of tropical species. Although estimates of global extinction rates are fraught with imprecision, species–area techniques do provide a means for shedding light on the probable effects of deforestation on species extinction. Based on species–area techniques and current rates of forest loss, over the next 25 years an estimated 4–8% of closed tropical forest species are likely to be committed to extinction. This loss will take place over a number of years as a new equilibrium number of species is achieved. If forest loss continues to accelerate, by the year 2040 some 17–35% of tropical forest species could be committed to eventual extinction when equilibrium numbers are reached. The 40–90% rise in the rate of deforestation of closed tropical forests that has occurred in the last decade has increased by 60–160% the number of tropical species likely to be condemned to extinction in the next half-century.

ACKNOWLEDGEMENTS

I thank Dan Janzen, Daniel Simberloff, and Mike Soulé for their helpful comments on a preliminary version of this manuscript, and Mohamed El-Ashry, Kenton Miller, Norman Myers, Jeff McNeely, and Peter Raven for their comments on more recent drafts.

REFERENCES

Connor, E.F. and McCoy, E.D, (1979) The statistics and biology of the species–area relationship. *American Naturalist*, **13**, 791–833.

Davis, S.D., Droop, S.J.M., Gregerson, P., Henson, L., Leon, C.J., Villa-Lobos, J.L., Synge, H. and Zantovska. J. (1986) *Plants in Danger: What Do We Know?* International Union for Conservation of Nature and Natural Resources, Gland, Switzerland.

FAO (Food and Agriculture Organization of the UN) (1981) *Tropical Forest Resources Assessment Project.* 4 vols. FAO, Rome.

FAO (1988) *An Interim Report on the State of Forest Resources in the Developing Countries.* FO:MISC/88/7, FAO, Rome.

FAO (1990) *Interim Report on Forest Resources Assessment 1990 Project.* Committee on Forestry, Tenth Session 24–28 September, COFO-90/8(a) Rome.

Fearnside, P.M. (1990) Deforestation in Brazilian Amazonia: the rates and causes of forest destruction (unpublished manuscript). National Secretariat of Science and Technology, National Institute for Research in Amazonia (INPA), Manaus, Brazil.

Forest Management Bureau (1988) *Natural Forest Resources of the Philippines.* Philippine German Forest Resources Inventory Project (pp. 62 mimeo).

Gentry, A.H. (1986) Endemism in tropical versus temperate plant communities, in *Conservation Biology: The Science of Scarcity and Diversity* (ed. M.E. Soulé), Sinauer Associates, Sunderland, MA, pp. 153–81.

Grainger, A. (1984) Quantifying changes in forest cover in the humid tropics: overcoming current limitations. *Journal of World Forest Resource Management*, **1**, 3–63.

Johns, A.D. (1986) *Effects of Habitat Disturbance on Rainforest Wildlife in Brazilian Amazon.* Final report, World Wildlife Fund US (WWF) project US-302. WWF, Washington, DC.

Johns, A.D. (1988) Economic development and wildlife conservation in Brazilian Amazonia. *Ambio*, **17**, 302–6.

Karr, J.R. (1982) Avian extinction on Barro Colorado Island, Panama: a reassessment. *American Naturalist*, **119**, 220–39.

Lanly, J.P. (1982) *Tropical Forest Resources*. Forestry Paper No. 30, Food and Agriculture Organization of the United Nations, Rome.

Lovejoy, T.E. (1980) A projection of species extinctions, in *Council on Environmental Quality* (CEQ), *The Global 2000 Report to the President*, Vol. 2. CEQ, Washington, DC, pp. 328–31.

Lugo, A.E. (1988) Estimating reductions in the diversity of tropical forest species, in *Biodiversity* (eds E.O. Wilson and F.M. Peter), National Academy Press, Washington, DC, pp. 58–70.

MacArthur, R.H. and Wilson, E.O. (1967) *The Theory of Island Biogeography*. Princeton University Press, Princeton, RI.

Malingreau, J.-P. and Tucker, C.J. (1988) Large-scale deforestation in the southeastern Amazon basin of Brazil. *Ambio*, **17**, 49–55.

Mares, M.A. (1986) Conservation in South America: Problems, consequences, and solutions. *Science*, **233**, 734–9.

Myers, N. (1979) *The Sinking Ark: A New Look at the Problem of Disappearing Species*. Pergamon Press, Oxford.

Myers, N. (1980) *Conversion of Tropical Moist Forests*. The National Research Council, National Academy of Sciences, Washington, DC.

Myers, N. (1988) Threatened biotas: 'hotspots' in tropical forests. *Environmentalist*, **8**(3), 1–20.

Myers, N. (1989) *Deforestation Rates in Tropical Forests and their Climatic Implications*. Friends of the Earth, London.

NRC (1980) *Research Priorities in Tropical Biology*. National Academy of Sciences, Washington, DC.

Ono, R.D., Williams, J.D. and Wagner, A. (1983) *Vanishing Fishes of North America*. Stone Wall Press, Inc., Washington, DC.

Quinn, J.F. and Harrison, S.P. (1988) Effects of habitat fragmentation and isolation on species richness: evidence from biogeographic patterns. *Oecologia*, **75**, 132–40.

Raup, D. M. (1978) Cohort analysis of generic survivorship. *Paleobiology*, **4**, 1–15.

Raven, P. H. (1987) The scope of the plant conservation problem world-wide, in *Botanic Gardens and the World Conservation Strategy* (eds D. Bramwell, O. Hamann, V. Heywood, and H. Synge), Academic Press, London, pp. 19–29.

Raven, P. H. (1988a) Biological resources and global stability, in *Evolution and Coadaptation in Biotic Communities* (eds S. Kawano, J.H. Connell, and T. Hidaka), University of Tokyo Press, Tokyo, pp. 3–27.

Raven, P.H. (1988b) Our diminishing tropical forests, in *Biodiversity* (eds E.O. Wilson and F.M. Peter), National Academy Press, Washington, DC, pp. 119–22.

Reid, W.V. and Miller, K.R. (1989) *Keeping Options Alive: The Scientific Basis for Conserving Biodiversity*. World Resources Institute, Washington, DC.

Sayer, J.A. and Whitmore, T.C. (1990) Tropical moist forests: destruction and species extinction. *Biological Conservation*, **55**, 199–201.

Shaffer, M.L. (1981) Minimum population sizes for species conservation. *BioScience*, **31**, 131–4.

Simberloff, D. (1986) Are we on the verge of a mass extinction in tropical rain forests? in *Dynamics of Extinction*, (ed. D.K. Elliott), Wiley, New York, NY, pp. 165–80.

Simberloff, D. and Abele, L.G. (1982) Refuge design and island biogeographic theory: effects of fragmentation. *American Naturalist*, **120**, 41–50.

Simon, J.L. (1986) Disappearing species, deforestation and data. *New Scientist*, 15 May, 60–3.

Sommer, A. (1976) Attempt at an assessment of the world's tropical moist forest. *Unasylva*, **28**(112 + 113), 5–24.

Wilson, E.O. (1988) The current state of biological diversity, in *Biodiversity* (eds E.O. Wilson and F.M. Peter), National Academy Press, Washington, DC, pp. 3–18.

Wilson, E.O. (1989) Threats to biodiversity. *Scientific American*, September 1990, 108–16.

Wong, M. (1985) Understorey birds as indicators of regeneration in a patch of selectively logged west Malaysian rainforest, in *Conservation of Tropical Forest Birds* (eds A.W. Diamond and T.E. Lovejoy), Technical Publication No. 4. International Council for Bird Preservation, Cambridge, UK, pp. 249–63.

WRI (World Resources Institute) (1990) *World Resources 1990–1991*. Oxford University Press, NY.

APPENDIX Deforestation of closed forests in tropical countries

	Closed forest in 1980s (000 ha)	Average annual deforestation in 1980s (000 ha)
AFRICA		
Angola	2 900	44
Benin	47	1
Cameroon	16 500	100
Central Afr. Rep.	3 590	5
Congo	21 340	22
Equatl. Guinea	1 295	3
Gabon	20 500	15
Ghana	1 718	22
Guinea	2 050	36
Guinea Bissau	660	17
Ivory Coast	4 458	250 (2)
Kenya	1 105	19
Liberia	2 000	46

APPENDIX *Cont.*

	Closed forest in 1980s (000 ha)	Average annual deforestation in 1980s (000 ha)
Madagascar	10 300	150
Nigeria	5 950	350 (3)
Reunion	82	na
Sénégal	220	na
Sierra Leone	740	6
Tanzania	1 440	10
Togo	304	2
Uganda	765	10
Zaïre	105 750	400 (2)
TOTAL AFRICA	203 714	1508
ASIA AND PACIFIC		
Australia[a]	600	na
Bangladesh	927	8
Brunei	323	5 (1)
Burma (Myanmar)	31 941	677
Fiji	811	2
Hawaii	445	na
India	36 540	1000 (3)
Indonesia	113 895	1000 (3)
Kampuchea	7 548	25
Laos	8 410	100
Malaysia	20 966	310 (3)
New Caledonia	484	na
Papua New Guinea	34 230	22
Philippines	9 510	143
Solomon Islands	2 423	1
Sri Lanka	1 659	58
Thailand	9 235	158 (3)
Vanuatu	236	4 (1)
Vietnam	8 770	200 (3)
TOTAL ASIA	288 953	3713
LATIN AMERICA		
Belize	1 354	9 (1)
Bolivia	44 010	87
Brazil	357 480	2666 (4)
Colombia	46 400	600 (3)
Costa Rica	1 638	124

APPENDIX *Cont.*

	Closed forest in 1980s (000 ha)	*Average annual deforestation in 1980s (000 ha)*
Cuba	1 455	2
Dominican Rep.	629	4
Ecuador	14 250	340
El Salvador	141	5
French Guiana	7 832	1 (1)
Guatemala	4 442	90
Guyana	18 475	2
Honduras	3 797	90
Jamaica	67	2
Mexico	46 250	700 (2)
Nicaragua	4 496	121
Panama	4 165	36
Peru	69 680	270
Puerto Rico	246	na
Surinam	14 873	3
Trinidad and Tobago	208	1
Venezuela	31 870	125
TOTAL LATIN AMERICA	673 715	5278
WORLD TOTAL	1 166 382	10 499

[a] Regional data were available to give a closer approximation to area of tropical closed forest (Grainger, 1984).

Source: WRI (1990), except: (1) FAO (1988), (2) Myers (1989), (3) WRI unpublished data (pers. comm. Norbert Henninger), (4) For August 1988 to August 1989, Fearnside *et al.* (1990).

from Whitmore, TC and JA Sayer, eds. 1992.
Tropical Deforestation and Species Extinction
Chapman & Hall, London.

— 4

Do species–area curves predict extinction in fragmented forest?

D. SIMBERLOFF

He made a game of elaborating the species of beetles to be found in a limited space. In Edward and Jiffy's garden alone, he found thirty-seven species. Within the range of a hundred feet of the hotel in any direction, he found more than fifty. If he extended his circle to a half-mile radius, his number of beetle species exceeded two hundred ... If fifty species could be found in an acre, and two hundred could be found in a square mile, Sewall estimated that half a million different species of beetle were lurking in the bushes within a day's ride of the hotel. (McMahon, 1979).

INTRODUCTION

Many forests throughout the world are undergoing severe reduction, but the loss is rarely simply the removal of contiguous areas. Instead, area loss is accompanied by more or less fragmentation. This process is not new, but attention was focused on it rather late. Curtis (1956) depicted the loss and fragmentation of forest in southern Wisconsin over a century and a half, and his picture was reprinted in many texts (e.g. Brown and Gibson, 1983) and even on a book cover (Burgess and Sharpe, 1981). An even longer record of fragmentation – 15 centuries – was published for the forests of Warwickshire (Thorpe, 1978), while Järvinen and Miettinen (1987) recently published a similar picture for two centuries of fragmentation of southern Finnish forests. Such maps are available for tropical forests, for example, five centuries of area reduction and fragmentation of forest in the vicinity of São Paulo.

No generally accepted theory predicts how many species might be lost to this simultaneous deforestation and fragmentation. Most research, including my own (Simberloff, 1986a), has focused on the species–area relationship. In its simplest statement, the species–area relationship is simply a monotonic increase in number of species with increase in area for a set of sites, all other things being equal. The problem is in the nature of that increase. Sewall in the epigraph assumed it to be linear, but few would support his conclusion that more than half a million beetle species inhabit a small part of Kansas. The species–area relationship is not really one theory; rather, it is an empirical generalization, perhaps the oldest in ecology; in the early nineteenth century, Watson (1835) described a version of the species–area relationship for British plants that seems to imply a power function. Since then, various theories have been proposed to account for the relationship.

THE SPECIES–AREA RELATIONSHIP

In a comprehensive review, Connor and McCoy (1979) examined 100 species–area relationships and drew two important conclusions. First, available data cannot distinguish between the two most widely proposed explanations. The first explanation is that area alone is important. This 'area *per se*' hypothesis is a consequence of the equilibrium theory of island biogeography (MacArthur and Wilson, 1967), which states that the number of species on an island is a dynamic equilibrium maintained by a balance between local extinction of species already on the island and immigration of species from source areas to the island. The rate of local extinction was envisioned as primarily determined by island area and what would nowadays be termed 'demographic stochasticity'. That is, the larger the island, the larger the population size for any particular species, on average, and thus the longer the expected time for any species to disappear from the island because of the sampling vagaries of birth and death among the finite number of individuals comprising the population. Thus, at any time, fewer species in the species pool would be (temporarily) extinct on a larger island than on a smaller one.

The second explanation for the species–area relationship states that larger sites, on average, have more habitats than small ones, and each habitat adds a number of species particularly adapted to it. Over large enough areas, there is no doubt that differing degrees of habitat diversity contribute to different numbers of species (e.g. Boecklen, 1986); habitat preferences of many species have been so well studied that they are part of the working knowledge of any ecologist. Species–area curves, though generally statistically significant, usually have much scatter. For example, among the 100 species–area relationships studied by Connor and McCoy (1979), the power function explained slightly less than half the variation in number of species (Boecklen and Gotelli, 1984). The remaining variation is often assumed to be largely caused by habitat differences among sites, although this hypothesis is not often tested. On the other hand, even over a range of very small areas a species–area relationship usually holds. For example, McGuinness (1984) depicts a generally monotonic relationship between number of plant and animal species and surface area of intertidal boulders with exposed surface areas between about 25 cm^2 and 1700 cm^2. Although distinct microhabitats are notoriously difficult to detect, one can certainly legitimately question whether there is more than one habitat if areas are small enough.

By contrast, over very small areas, there is direct evidence in some systems for occasional local extinction of species (e.g. Simberloff, 1976), the force behind the area *per se* hypothesis. But there has been much debate over whether extinctions occur with sufficient frequency on large areas to play a key role in determining number of species (e.g. Gilbert, 1980; Simberloff, 1983). For some taxa the same contention has been raised even for very small areas (e.g. Estabrook, 1990). Of course, it need not be true that only one force acts to produce all species–area relationships, or even one species–area relationship. Williams (1964) suggested that different forces typically act over different ranges of area to produce the monotonic relationship.

Connor and McCoy (1979) also determined that no one canonical model fits all species–area relationships best. In the wake of Preston's (1962) attempt to derive a particular form of the species–area relationship from his conception of a canonical lognormal species-abundance curve, and MacArthur and Wilson's (1967) popularization of this approach, it came to be common practice to represent species–area relationships as a power function:

$$S = kA^z \quad \text{or} \quad \log S = k + z \log A$$

This shape was first proposed by Arrhenius (1921). Gleason (1922), observing that Arrhenius's equation gave excessive estimates of species richness when extrapolated to large areas, proposed an exponential relationship:

$$S = \log k + z \log A$$

Until Preston's work, the exponential model received as much attention as the power model, but the power model is nearly universal nowadays. This is in spite of the fact that it is not the best model for many data sets. Connor and McCoy (1979) found that 35 of the 100 data sets were best fit by a relationship in which neither variable is transformed, and only 36 of the remaining 65 were best fit by the power function. Connor and McCoy did not attempt various curvilinear relationships, but often other relationships fit better than the untransformed model, the exponential model, and the power model. For example, Simberloff and Gotelli (1984) found for plant species in prairie remnants a better fit by:

$$S^{1/2} = a + b \log A$$

As there is no clear best-fit, standard model of the species–area relationship, it is not surprising that Connor and McCoy found no ready biological interpretation for the parameters of the power function. Others disagree and find particular interpretations for z, but attempting to calculate z-values and to predict species richness for a given area using z-values is, at best, problematic (Martin, 1981; Lomolino, 1989). In particular, the confidence bands around regression-fitted species–area curves are notoriously wide, particularly outside the domain of the observed points, and parameter estimates are very sensitive to particular influential data points (Boecklen and Gotelli, 1984; Boecklen and Simberloff, 1986).

Nevertheless, it is a worthwhile exercise to use the species–area relationship to attempt a first guess at how many extinctions deforestation will generate in tropical forests (e.g. Simberloff 1986a; Reid, Chapter 3). The general strategy is to assume some form of the species–area relationship, then produce an equation for the fraction of species remaining for a particular fraction of area remaining. For example, if one assumes the power function, then

$$S_{original} = c(A_{original})z$$

$$S_{new} = c(A_{new})z$$

So

$$S_{new}/S_{original} = (A_{new}/A_{original})z$$

In order to see the limitations of this approach, consider a concrete case. Suppose a forest is to be reduced to 20% of its area, and one has somehow fitted a power function to similar forests (this is not a trivial matter; see Connor and McCoy, 1979). Values of the exponent in such an exercise are usually in the range of 0.15–0.4 (Connor and McCoy, 1979). It makes a very big difference whether z is, for example, 0.2 or 0.3. In the first case, at equilibrium 0.725 of the original species number would be predicted, whereas in the second case, 0.617 would be predicted. The enormous confidence limits noted above should make one view any such prediction with great caution, particularly because the original species–area relationship is unlikely to have been calculated from sample points in the range of the new area. Thus, the confidence limits will be particularly wide and, as noted above, one might not even expect the same forces to be determining the species–area relationship.

In any event, there is no accepted theory for the rate at which species will be lost through time and the new equilibrium achieved, even if one were to accept the estimation problems noted above (Simberloff, 1986a).

The closest analogue in the ecological literature is various studies of the decline of species richness (*viz* 'relaxation') with changed area (e.g. Diamond, 1972; Terborgh, 1974) and in refuges (Soule *et al.*, 1979). However, these efforts are so fraught with untenable assumptions and statistical errors (Abele and Connor, 1979; Faeth and Connor, 1979; Boecklen and Gotelli, 1984; Boecklen and Simberloff, 1986) that they are simply not useful as either empirical analogues or guides to the development of relevant theory.

FRAGMENTATION

Aside from these statistical problems, there are two serious difficulties in using the species–area relationship to predict the effect of current and projected patterns of deforestation. Using the species–area relationship, with all its warts, in such an endeavour entails the assumption that a single uniform-habitat forest with species spread uniformly throughout it is reduced to another single uniform-habitat forest with species spread uniformly throughout it. However, all forests are patchwork quilts of various habitats, so that species are not uniformly distributed. Further, various historical factors may have led certain parts of a large forested region to be particularly species-rich. For example, certain areas appear to be 'hot-spots' of endemism (Myers, 1990a). And, as noted at the outset, the forest is fragmented in addition to being reduced.

The habitat heterogeneity and other ways in which certain areas come to have vastly more species than others could be viewed as cause for both optimism and pessimism. On the one hand, a judicious choice of habitats could greatly increase the number of species that might be saved. For example, Kitchener *et al.* (1980) suggested that almost all lizard species of Australia's wheatbelt could be maintained in a series of small, scattered refuges selected for their habitats, whereas a single contiguous area would have to be hundreds of times larger to serve the same purpose. On the other hand, at least in tropical forest, areas of particularly high numbers of species seem to be particularly threatened (Raven, 1987; Myers, 1990b), so that more species will be lost than a simple species–area calculation would have predicted. Minimizing extinction by dealing with spatial heterogeneity in species richness would entail research of a sort already understood, even though it is not being conducted at a nearly sufficient rate. What is required is autecological studies of habitat requirements of species (many of which are not even known) and biogeographical and ecological analyses of spatial variation of diversity. However, the nature of the challenge to understanding the impact of fragmentation seems unappreciated and, to some extent, misunderstood. Thus the remainder of this chapter will concentrate on fragmentation.

All the species–area relationship predicts about the number of species in an archipelago of forest islands in a sea of cut-over land is that each fragment will have fewer species than the entire original. It says nothing whatever about how many species the entire archipelago will have relative to the original. This is a matter of the degree of overlap among sets of species in the different forest islands (Simberloff and Abele, 1976, 1982).

This question at first seems analogous to the SLOSS (single-large-or-several-small) issue that, for a time in the 1980s, rent some segments of the academic community. The question was, should one aim to have a single large or several small refuges, of equal total area? Although, if one's aim is solely to conserve the greatest number of species (usually *not* the goal of refuge design), the answer to the SLOSS question is relevant to the question of how many species will be extinguished by tropical deforestation, it is far from providing a complete answer. This is because the refuges envisioned in the SLOSS debate are generally all very small fragments of an originally continuous habitat, and the loss of area alone will almost certainly account for more of the extinction in the refuge system than will the particular configuration of the component refuges. This book is concerned with the total loss of diversity, whereas the SLOSS debate largely takes a major loss of diversity as a given (and an unknown) and concerns itself with maximizing the remaining diversity.

Similarly, many patch dynamics models (references in Quinn and Robinson, 1987; Robinson and Quinn, 1988; Simberloff, 1988) predict lower aggregate extinction rates and higher diversity for an ensemble of small sites compared to a single larger one, but again, these models do not address the conservation implications of tropical deforestation. They may be relevant to persistence of diversity in one ensemble of small refuges as opposed to another, but they do not treat the major loss of area that would precede the patchy condition that they take as an assumption.

The main empirical approach to the SLOSS debate reflects this perspective. Many researchers looked at archipelagoes of habitat fragments, comparing species sets on single larger fragments to the combined species sets of groups of smaller fragments, with total area held constant (references in Simberloff and Abele, 1982, Simberloff and Gotelli, 1984). The usual result was that the collections of smaller fragments had at least a few more species than the single larger one. This observation might lead to optimism about fragmentation, but this type of research in fact has limited implications for predictions of number of extinctions that will accompany tropical deforestation.

One reason is that, as in the argument about the SLOSS issue in refuge design, these fragments were already a small fraction of the original area, and, even aside from problems of habitat heterogeneity, it is difficult to extrapolate statistically from the number of species in small samples to the number of species in the universe as a whole (e.g. Lewins and Joanes, 1984; Smith and van Belle, 1984). Thus, it is difficult to estimate the number of species originally present.

A second reason is that, if fragments are made small enough, they will all be too small for a large fraction of the species. The critical size must be determined empirically, but the principle is well established. Every species has some minimum viable population size (or better, range of sizes) below which swift extinction is very likely (Shaffer, 1981; Simberloff, 1988). Small populations are likely to disappear for many reasons. Which ones are important for which species, and at what population size they become important, can be determined only by studying particular species. Among factors suggested for particular cases are demographic stochasticity and inbreeding depression. Several factors may interact to doom very small populations (Simberloff, 1986b; Gilpin and Soule, 1986). The important point is not so much what forces produce what precise threshold as that the threshold exists. As the area of a fragment gets smaller and smaller, it becomes too small to maintain minimum viable population sizes of more and more species. Having even thousands of small fragments, with large aggregate area, will not be expected to allow conservation of many species, because all will be too small and isolated for most species. This phenomenon is above and beyond whatever habitat modification will occur as fragments are increasingly reduced. For example, smaller fragments contain proportionally more edge habitat.

As area reduction and fragmentation proceed, extinction will come in two broad forms. First, some species will disappear as their habitat is completely destroyed within their range before they disperse or someone moves them. The number of species dying in this way cannot be estimated because species ranges are poorly known. Given that most species in tropical forests are insects, the majority of tropical species are not even described (e.g. Erwin, 1988). However, the number that will be lost this way will probably be substantial. The species going extinct by this route without our knowledge must greatly exceed those that we do know about because one biological trait predisposing a species to be undescribed is small geographic range, and this same feature would predispose a species to extinction by having its habitat completely destroyed within its range (Simberloff, 1984). Apparently tropical species' ranges are often very small, probably on average much smaller than those of temperate species. Neotropical bird species' ranges are smaller than nearctic bird species' ranges, on average (Terborgh and Winter, 1982), but I know of no similar analysis of other taxa. However, one emphasis in the tropical literature is on the great proportion of species that appear to have very narrow ranges (e.g. Erwin, 1988; Gentry, 1989).

Other species will not be eliminated quickly, either because their habitat is not completely destroyed within their range or they migrate to new areas of suitable habitat as their old areas are destroyed. Some genetic variation may be lost as distinct local genotypes disappear, but the species will persist. They will then be analogous to species isolated in refugia by various geological events, for example, Pleistocene temperature changes. Some will dwindle to extinction in the remaining fragments, as described above, if their population sizes fall below the minimum viable population size. As noted, how quickly this occurs for each species, and how quickly the total species number declines, cannot currently be forecast by any theory and the empirical data are too meagre to be used for prediction.

One way to approach this lacuna in our knowledge would be direct controlled experiments on fragmentation, seeking both the specific reasons for individual species decline and the rate of loss of species. However, the number of such experiments known to me is pitifully small, and only two of those (one in tropical forest) seem to entail a relevant range of fragment sizes. In California grassland (Quinn and Robinson, 1987; Robinson and Quinn, 1988) and Kansas prairie (R. Holt, *pers. comm.*) different degrees of fragmentation produced archipelagoes of different sizes and numbers of fragments, but total archipelago area was but a fraction of a hectare and fragments were only a few metres apart. Although such studies can provide insight into the behaviour of particular species in such a landscape, it would be rash to extrapolate results from such studies to different species in a different habitat with fragments that would range from perhaps a hectare to thousands of hectares.

The Minimum Critical Size of Ecosystem project of the World Wildlife Fund in Brazilian Amazonian rain forest (Lovejoy *et al.*, 1983) is another matter entirely. Fragments range from 1 ha to 10 000 ha and patterns are already evident in the decline of frogs (Zimmerman and Bierregaard, 1986) and birds (R. Bierregaard, *pers. comm.*). The particular mechanisms responsible for extinctions, though idiosyncratic, suggest that species–area relationships might not capture major processes leading to extinction in fragmented as opposed to intact landscapes with equal total forested area. Most understorey birds will not willingly cross gaps of even 80 m. The bat falcon, *Falco rufigularis*, has repeatedly been seen to chase birds flying over cleared areas from fragment to fragment. Certain frog species missing from particular fragments, by contrast, appear to be absent because their breeding habitat is absent. They require small permanent muddy ponds, which are usually maintained in this region by pigs. A fragment without pigs will lack these frogs. There is no immediate area limitation in the smaller fragments; as long as pigs wallow in them, they suffice for the frogs. Of course, if the probability of having a pig wallow depended on area, the habitat limitation would be somewhat reflected in an area relationship.

Another controlled experiment with fragments of substantial area is that conducted by David Lindenmayer of the Victoria Department of Conservation and Environment in temperate rain forest in Australia. This project is new, but should yield results at least on birds and mammals comparable to the Brazilian experiment.

A handful of other studies, though not producing fragments in a controlled, replicated way and watching what happens in them, have focused on the specific mechanisms by which particular systems and species respond to fragmentation. None of these is in tropical forest, but to my mind they constitute the most important kind of research being conducted by conservation biologists. Together, they suggest that fragmentation, independently of area loss, will lead to extinction of many species.

Consider the current *cause célèbre* of United States conservation, the northern spotted owl, *Strix occidentalis caurina* (Simberloff, 1987). The owl is highly associated with old-growth rain forests of the Northwest, where perhaps 2000 pairs are left. As this habitat has declined from about 7.5 million ha to at most one million ha (Morrison, 1988), the owl has been increasingly restricted to fragments in a patchy landscape. Most attention has been paid to the enormous area requirements for this species to find enough prey (predominantly rodents) to reproduce and fledge young. A typical home range for a pair varies from perhaps 1500 ha in the southern part of its distribution to over 3000 ha in the northern part.

However, recent findings about yearling dispersal mortality in the current fragmented landscape imply that, even if home ranges of this size were provided, the bird would still be doomed if these were scattered. As many as 80% of yearling males die, apparently from predation by great horned owls (*Bubo virginianus*) and goshawks (*Accipiter gentilis*) as they disperse over cleared areas (R. Gutierrez, *pers. comm.*). Before deforestation they rarely had to leave closed forest. This finding helped persuade an interagency federal committee charged with producing a management plan for the owl to recommend a shift in strategy, from SOHAs (spotted owl habitat areas) of sizes towards the low end of observed home ranges to HCAs (habitat conservation areas) that consist of much larger, contiguous blocks (Thomas *et al.*, 1990).

Nest predation in the fragmented eastern forest of the United States may contribute to the decline of some songbirds. Wilcove (1985) placed artificial nests with quail eggs at various distances from the edge within forest patches of various sizes. He found a strong area relationship between egg predation and both size and distance from edge. For example, in the largest patch (200 000 ha), only 2% of the nests were preyed upon within a week. In 4–12 ha suburban woodlots, predation was 70%. Wilcove suggested that most predation was probably by intermediate-sized species that would otherwise have been held in check by larger predators, now absent from the fragmented landscape. However, further empirical work is required to test this hypothesis.

Upland forests of northern Wisconsin originally consisted of perhaps 80% old growth with the remaining 20% as small islands of early successional stages. The pattern is now the opposite, with small islands (8–200 ha) of old growth embedded in a sea of early successional species, particularly aspen (*Populus tremuloides*). White-tailed deer (*Odocoileus virginianus*) populations have more than doubled as logging has generated a mosaic of clearcuts and secondary successional areas with abundant deer browse. The enhanced browsing selects against key original groundcover plants and also seedlings of typical old growth dominant tree species. Thus Alverson *et al.* (1988) argue that, even if the US Forest Service planned to maintain many of these small old-growth islands, these patches cannot persist. Rather, they argue that, if old growth is to be maintained, it will have to be in much larger stands.

Forests dominated by longleaf pine (*Pinus palustris*) once covered about 28 million hectares of the south-eastern United States, largely in continuous stands. No more than 1000 ha of the virgin forest remains, and, of some four million ha of second growth, most lacks the original ground cover. This forest is home to the endangered red-cockaded woodpecker (*Picoides borealis*) and many other rare animals. The woodpecker requires large, old, diseased pines, in stands of over 50 ha, for nesting and rearing its young (Wood, 1983; Jackson, 1988). Longleaf pine forests are maintained by frequent fires; otherwise they are succeeded by hardwood forests (Schiff, 1962). To carry the fire, good ground cover is needed; this is why much of the second growth does not behave like primary longleaf pine forest. Lightning starts the fires and, when forests were continuous, most longleaf areas burned about every three years; even if lightning did not strike a particular area, it would strike nearby and the fire would spread. Now the tiny longleaf fragments are separated from one another and fires naturally occur in each of them rarely. This would be true even if there were many more longleaf fragments, so long as they were isolated fragments. For now, a costly stewardship programme is required to maintain this entire community, simply by virtue of its fragmented state.

The Rollins Sanctuary, a 50 ha tropical hardwood forest on Fort George Island, part of metropolitan Jacksonville, Florida, is intended to be maintained in perpetuity. Short of regional climatic change, this might not have been a problem when all Fort George Island (375 ha) was tropical hardwood forest. Now, however, like much of surrounding Jacksonville, it is a sea of exotic and ornamental plants. It seems likely that seeds of these introduced species will compete strongly with seeds produced by trees within the sanctuary. Even a network of such forests, with substantial total area, would probably by threatened by a rain of exotic propagules.

Finally, the examples just cited focus on the biology of particular species. Fragmentation also induces change in a number of physical factors, such as radiation, wind, and water flux. Although the general impact on species and communities of such microclimatic changes has not been well studied in many systems, the changes themselves have been so well documented and impacts on particular species in some systems seem so apparent and suggestive (Saunders *et al.*, 1991) that it would be foolish to discount the effect of such changes on future species persistence.

CONCLUSIONS

The above examples all described specific responses of particular species or communities to fragmentation. Although these responses are idiosyncratic, already one problem has cropped up repeatedly even in the very few intensively studied systems. Increased predation in a fragmented landscape is a common theme; it occurs in the tropical forest example in Brazil. Another problem, that of increased access of exotic species in a fragmented landscape, will almost certainly be generic. Possibly the most important implication, however, stems from the variety of ways in which fragmentation has been shown to be inimical in so few studies. None of these cases would likely by manifested if the remaining area were in one contiguous mass, and the mechanisms for these problems are not those envisioned by either of the two explanations of the species–area relationship – the area *per se* hypothesis and the habitat diversity hypothesis. There is every reason to think that careful study would reveal myriad similar problems. Probably any species that has evolved in large, relatively continuous habitat has traits that are maladaptive in small, isolated fragments. Thus, even the substantial number of extinctions predicted from species–area relationships will probably severely underestimate the extinctions that will actually occur.

It is a commonplace that forests of the eastern United States were reduced over two centuries to fragments totalling 1–2% of their original extent, and that during this destruction, only three forest birds went extinct – the Carolina parakeet (*Conuropsis carolinensis*), the ivory-billed woodpecker (*Campephilus principalis principalis*), and the passenger pigeon (*Ectopistes migratorius*). Although deforestation certainly contributed to the decline of all three species, it was probably not critical for the pigeon or the parakeet (Greenway, 1967). Why, then, would one predict massive extinction from similar destruction of tropical forest?

The answer to this question relates to two points raised earlier. First, species–area relationships are expected to obtain at equilibrium. As noted above, no theory credibly predicts how long it will take for extinction to produce this equilibrium. For long-lived species, one might expect it to take several centuries. It is possible that the stage has already been set for substantial extinctions of some taxa in the eastern United States. Second, the large geographic ranges of nearctic birds greatly aided them to avoid extinction from habitat destruction; neotropical bird species' ranges are, on average, much smaller (Terborgh and Winter, 1982). A related important point is that, although more is known about bird extinctions than about those for any other group except possibly mammals, birds are likely to be the poorest possible indicator taxon for extinction of other species. Largely because of their dispersal ability, they tend to have ranges larger than those of other species. Furthermore, the same dispersal ability aids them to move to suitable new habitat when their old habitat is destroyed. Many neotropical cloud forest plant species are endemic to isolated sites smaller than 10 km^2 (Gentry, 1989). Thus, extinction rates of nearctic birds subject to deforestation and fragmentation will probably not predict those even for tropical bird species, much less for other tropical species.

SUMMARY

Fragmentation of forests threatens the persistence of some species independently of the species loss predicted by the species–area relationship. No existing theory adequately predicts the extent of extinction that will be caused by fragmentation and loss of area acting jointly. Similarly, for neither sort of species loss can the time course of extinction be predicted. Empirical studies in a variety of fragmented forests suggest that effects may be major but are likely to be idiosyncratic. For example, increased herbivory or predation of particular species may increase in a fragmented landscape. If the prey or host plant is a species affecting many others in the community, the entire nature of the community may change. Similarly, successional processes may be impeded by fragmentation and lead to a different sort of climax community. It is likely that fragmentation will have greater effects in the tropics than elsewhere, but adequate empirical data are lacking.

REFERENCES

Abele, L.G. and Connor, E.F. (1979) Application of island biogeography theory to refuge design: making the right decision for the wrong reasons, in *Proceedings of the First Conference on Scientific Research in the National Parks*, Vol. 1, (ed. R.M. Linn), USDI, Washington, pp. 89–94.

Alverson, W.S., Waller, D.M. and Solheim, S.L. (1988) Forests too dear: edge effects in northern Wisconsin. *Conservation Biology*, 2, 348–58.

Arrhenius, O. (1921) Species and area. *Journal of Ecology*, **9**, 95–9.

Boecklen, W.J. (1986) Effects of habitat heterogeneity on the species–area relationship of forest birds. *Journal of Biogeography*, **13**, 59–68.

Boecklen, W.J. and Gotelli, N.J. (1984) Island biogeographic theory and conservation practice: species–area or specious–area relationships? *Biological Conservation*, **29**, 63–80.

Boecklen, W.J. and Simberloff, D. (1986) Area-based extinction models in conservation, in *Dynamics of Extinction*, (ed. D.K. Elliott), Wiley, New York, pp. 247–76.

Brown, J.H. and Gibson, A.C. (1983) *Biogeography*. C.V. Mosby, St Louis.

Burgess, R.L. and Sharpe, D.M. (1981) *Forest Island Dynamics in Man-Dominated Landscapes*. Springer-Verlag, New York.

Connor, E.F. and McCoy, E.D. (1979) The statistics and biology of the species–area relationship. *American Naturalist*, **113**, 791–833.

Curtis, J.T. (1956) The modification of mid-latitude grasslands and forests by man, in *Man's Role in Changing the Face of the Earth* (ed. W.L. Thomas), University of Chicago Press, Chicago, pp. 721–36.

Diamond, J.M. (1972) Biogeographic kinetics: estimation of relaxation times for avifaunas of Southwest Pacific Islands. *Proceedings of the National Academy of Sciences USA*, **69**, 3199–203.

Erwin, T.L. (1988) The tropical forest canopy. The heart of biotic diversity, in *Biodiversity* (eds E.O. Wilson and F.M. Peter), National Academy Press, Washington, pp. 123–9.

Estabrook, G.F. (1990) The size of nature reserves and the number of long lived plant species they contain. *Coenoses*, in press.

Faeth, S.H. and Connor, E.F. (1979) Supersaturated and relaxing island faunas: a critique of the species–area relationship. *Journal of Biogeography*, **6**, 311–16.

Gentry, A.H. (1989) Speciation in tropical forests, in *Tropical Forests, Botanical Dynamics, Speciation and Diversity* (eds L.B. Holm-Nielson, I.C. Nielson, and H. Balshev), Academic Press, San Diego, pp. 113–34.

Gilbert, F.S. (1980) The equilibrium theory of island biogeography: fact or fiction? *Journal of Biogeography*, **7**, 209–35.

Gilpin, M.E. and Soule M.E. (1986) Minimum viable populations: processes of species extinction, in *Conservation Biology: The Science of Scarcity and Diversity* (ed. M.E. Soule), Sinauer, Sunderland, MA, pp. 19–34.

Gleason, H.A. (1922) On the relation between species and area. *Ecology*, **3**, 158–62.

Greenway, J.C. Jr (1967) *Extinct and Vanishing Birds of the World*. Dover, New York.

Jackson, J.A. (1988) The southeastern pine forest ecosystem and its birds: past, present, future, in *Bird Conservation* vol.3 (ed. J.A. Jackson), University of Wisconsin Press, Madison, WI, pp. 119–59.

Järvinen, O. and Miettinen, K. (1987) *Sammuuko Suuri Suku?* Suomen Luonnonsuojelun Tuki Oy, Helsinki.

Kitchener, D.J., Chapman, A., Dell, J., Muir, B.G. and Palmer, M. (1980) Lizard assemblage and reserve size and structure in the Western Australian wheatbelt – some implications for conservation. *Biological Conservation*, **17**, 25–62.

Lewins, W.A. and Joanes, D.N. (1984) Bayesian estimation of the number of species. *Biometrics*, **40**, 323–8.

Lomolino, M.V. (1989) Interpretations and comparisons of constants in the species–area relationship: an additional caution. *American Naturalist*, **113**, 277–80.

Lovejoy, T.E., Bierregaard, R.O., Rankin, J.M. and Schubart, H.O.R. (1983) Ecological dynamics of tropical forest fragments, in *Tropical Rain Forest: Ecology and Management* (eds S.L. Sutton, T.C. Whitmore, and A.C. Chadwick), Blackwell, Oxford, pp. 377–84.

MacArthur, R.H. and Wilson, E.O. (1967) *The Theory of Island Biogeography*. Princeton University Press, Princeton, New Jersey.

Martin, T.E. (1981) Species–area slopes and coefficients: a caution on their interpretation. *American Naturalist*, **118**, 823–37.

McGuinness, K.A. (1984) Species–area relations of communities on intertidal boulders: testing the null hypothesis. *Journal of Biogeography*, **11**, 439–56.

McMahon, T. (1979) *McKay's Bees*. Harper and Row, New York.

Morrison, P.H. (1988) *Old Growth in the Pacific Northwest: A Status Report*. The Wilderness Society, Washington.

Myers, N. (1990a) The biodiversity challenge: expanded hot-spots analysis. *The Environmentalist*, **10**, 1–14.

Myers, N. (1990b) Mass extinctions: what can the past tell us about the present and the future? *Palaeogeography, Palaeoclimatology, Palaeoecology*, **82**, 175–85.

Preston, F.W. (1962) The canonical distribution of commonness and rarity. *Ecology*, **43**, 185–215, 410–32.

Quinn, J.F. and Robinson, G.R. (1987) The effects of experimental subdivision on flowering plant diversity in a California annual grassland. *Journal of Ecology*, **75**, 837–56.

Raven, P.H. (1987). We're killing our world: the global ecosystem in crisis. MacArthur Foundation Occasional Paper, MacArthur Foundation, Chicago, IL.

Robinson, G.R. and Quinn, J.F. (1988) Extinction, turnover and species diversity in an experimentally fragmented California annual grassland. *Oecologia*, **76**, 71–82.

Saunders, D.A., Hobbs, R.J. and Margules, C.R. (1991) Biological consequences of fragmentation: A review. *Conservation Biology*, **5**, 18–32.

Schiff, A.L. (1962) *Fire and Water – Scientific Heresy in the Forest Service*. Harvard University Press, Cambridge, MA.

Shaffer, M.L. (1981) Minimum viable population sizes for species conservation. *BioScience*, **31**, 131–4.

Simberloff, D. (1976) Species turnover and equilibrium island biogeography. *Science*, **194**, 572–78.

Simberloff, D. (1983) Biogeography: the unification and maturation of a science, in *Perspectives in Ornithology: Essays Presented for the Centennial of the American Ornithologists' Union* (eds A.H. Brush and G.A. Clark, Jr), Cambridge University Press, Cambridge, pp. 411–55.

Simberloff, D. (1984) Mass extinction and the destruction of moist tropical forests. *Zhurnal Obshchei Biologii (Journal of General Biology) USSR*, **45**, 767–78.

Simberloff, D. (1986a) Are we on the verge of a mass extinction in tropical rain forests? in *Dynamics of Extinction* (ed. D.K. Elliott), Wiley, New York, pp. 165–80.

Simberloff, D. (1986b) The proximate causes of extinction, in *Patterns and Processes in the History of Life*, (eds D.M. Raup and D. Jablonski), Springer-Verlag, Berlin, pp. 259–76.

Simberloff, D. (1987) The spotted owl fracas: mixing academic, applied and political ecology. *Ecology*, **68**, 766–82.

Simberloff, D. (1988) The contribution of population and community biology to conservation science. *Annual Review of Ecology and Systematics*, **19**, 473–511.

Simberloff, D. and Abele, L.G. (1976) Island biogeographic theory and conservation practice. *Science*, **191**, 285–6.

Simberloff, D. and Abele, L.G. (1982) Refuge design and island biogeographic theory: Effects of fragmentation. *American Naturalist*, **120**, 41–50.

Simberloff, D. and Gotelli, N.J. (1984) Effects of insularisation on plant species richness in the prairie-forest ecotone. *Biological Conservation*, **27**, 46.

Smith, E.P. and van Belle, G. (1984) Nonparametric estimation of species richness. *Biometrics*, **40**, 119–29.

Soule, M.E., Wilcox, B.A. and Holtby, C. (1979) Benign neglect: a model for faunal collapse in the game reserves of East Africa. *Biological Conservation*, **15**, 259–72.

Terborgh, J. (1974) Preservation of natural diversity: the problem of extinction-prone species. *BioScience*, **24**, 715–22.

Terborgh, J. and Winter, B. (1982) Evolutionary circumstances of species with small ranges, in *Biological Diversification in the Tropics* (ed. G.T. Prance), Columbia University Press, New York, pp. 587–600.

Thomas, J.S., Forsman, E.D., Lint, J.B., Meslow, E.C., Noon, B.R. and Verner, J. (1990) *A Conservation Strategy for the Northern Spotted Owl*. USDA Forest Service, USDI Bureau of Land Management, USDI Fish and Wildlife Service, and USDI National Park Service, Portland, Oregon.

Thorpe, H. (1978) The man–land relationship through time, in *Conservation and Agriculture* (ed. J.G. Hawkes), Duckworth, London, pp. 17–44.

Watson, H.C. (1835) Remarks on the geographical distribution of British plants. London, n.p.

Wilcove, D. (1985) Nest predation in forest tracts and the decline of migratory songbirds. *Ecology*, **66**, 1211–14.

Williams, C.B. (1964) *Patterns in the Balance of Nature and Related Problems in Quantitative Ecology*. Academic Press, London.

Wood, D.A. (ed.) (1983) *Proceedings of the Red-cockaded Woodpecker Symposium*. Florida Game and Fresh Water Fish Commission, Tallahassee, Fla.

Zimmerman, B.L. and Bierregaard, R.O. (1986) Relevance of the equilibrium theory of island biogeography and species–area relations to conservation with a case from Amazonia. *Journal of Biogeography*, **13**, 133–43.

[handwritten margin note: tropical vs. temperate range sizes? (p. 81)]

*from Whitmore, TC and JA Sayer, eds. 1992.
Tropical Deforestation and Species Extinction
– Chapman & Hall, London –*

— 5

Species extinctions in tropical forests

V.H. HEYWOOD and S.N. STUART

*nothing on ecosystem
service aspects of
biodiversity*

INTRODUCTION

Most discussions on extinction rates in tropical forests have tended to focus on predictions of the large numbers of species that are likely to become extinct within the next 10–50 years. Attention is usually drawn to the probable losses that can be expected in areas of high biological diversity and endemism, such as the Atlantic Forest of Brazil (Brown, Chapter 6), or the forests of Madagascar, which have suffered contractions of anything up to 90% or more of their original area in the last 40–50 years (cf. Myers, 1986, 1988a; Wilson, 1988). In such areas, in accordance with the species–area relationship which has been used to predict extinctions (cf. Reid, Chapter 3), it has been suggested that 50% of the plant and animal species are at risk of imminent extinction, assuming they have not already disappeared. The principal reason for this predicted loss is the continued conversion of forests to other uses, largely as a result of pressures caused by rapidly growing human populations.

Raven (1990) expresses the situation as follows:

> The regions where tropical moist forests are
> expected to be destroyed in the next 30 years
> contain an estimated half of the world's plants (. . .).
> If half of these species will be at risk
> when the forests have been reduced to less
> than 10% of their original extent, I
> estimate the extinction of perhaps 65 000
> species, a quarter of the world's total,
> within the next several decades.

The Joint IUCN-WWF Plants Programme suggested that if present trends of habitat loss continue, up to 60 000 plant species may become extinct *or suffer severe genetic erosion* (our emphasis) by the middle of the next century. Such loss of genetic variation in species' populations is the very essence of the process of extinction, and this concern is echoed by Ehrlich (1988) who notes: 'The loss of genetically distinct populations within species is, at the moment, at least as important a problem as the loss of entire species. Once a species is reduced to a remnant, its ability to benefit humanity ordinarily declines greatly and its total extinction in the relatively near future becomes much more likely'. The genetic resources sector has fully recognized this problem, although its full extent can only be guessed at.

The formulation by the IUCN-WWF Plants Programme, which embraces both species falling below a specified level of abundance and those suffering total extinction is referred to by Ginzberg *et al.* (1982) as 'quasiextinction'. As Ginzberg (1990) notes, the concept of quasiextinction is a two-edged sword: on the one hand it admits the native stochasticity that is inherent in ecological processes, but on the other hand suffers from the difficulty that it can lead to probabilistic predictions which are hard to falsify. But as Ginzberg *et al.* (1982) say, a practical approach to this difficulty is to evaluate scrupulously the assumptions of the models used to make such estimations. We suggest applying a similar rigour in evaluating the predictions of extinction rates. As Wilson (1988) comments, it is curious that the study of extinctions remains one of the most neglected areas in ecology, and there is a pressing need, he says, for a more sophisticated body of theories and carefully planned field studies on it than now exist. We agree.

The purpose of this chapter is to:

1. Examine the field evidence for mass extinctions taking place at the present time.
2. Assess the extent to which large numbers of species might already be committed to extinction.
3. Examine the problems involved in predicting species loss.
4. Assess what effects conservation action might have had on extinction rates.
5. Outline the practical conservation actions that can be taken to reduce species loss and genetic erosion.

It is not the main purpose of this chapter to challenge the various methodologies employed to calculate extinctions rates, although we do draw attention to problems posed by some of the assumptions that they involve. We also scrutinize some of the statements that have been made concerning extinctions by recent authors as it is not at all clear, in many cases, exactly what it is they are purporting to say. For example, when deforestation reaches 90% of an original area, is it assumed that this will lead to the *immediate* or *imminent* extinction of up to 50% of the species, has *already* led to the extinction of up to 50% of the species or has *committed* up to 50% of the species to extinction at some future date? We attempt to shed light on these issues in this chapter.

We do not question current estimates of extinction as though the matter were one of just academic concern. We are concerned with the practicalities of conservation action that may serve to counter the effects of habitat loss and the risks of species extinctions. Also, whatever the level of accuracy or probability of these predictions, we recognize that they have served the important purpose of advocacy with respect to the environment. This is put well by Lovejoy (1989) who writes that if science does not take on such a role, 'we deserve and can expect the future censure of society, for indeed it is our responsibility, as those who understand best what is happening and what alternatives exist, to sound the tocsin about environmental deterioration and conservation problems in all their variety.' He also recommends that we need to build a margin of error into our recommendations to hedge our bets against our current ignorance. We accept this moral responsibility and also the philosophy of what may be termed the precautionary principle – the desirability of risking taking too much action rather than too little. However, we also believe that it is essential to ensure that our statements are scientifically defensible, since too large margins of error could undermine the credibility of the conservation case in the political arena. This is a delicate balance to achieve, and we attempt to walk the tightrope in this chapter.

FIELD EVIDENCE FOR MASS EXTINCTIONS

IUCN, together with the World Conservation Monitoring Centre, has amassed large volumes of data from specialists around the world relating to species decline, and it would seem sensible to compare these more empirical data with the global extinction estimates. In fact, these and other data indicate that the number of recorded extinctions for both plants and animals is very small and so most of the evidence for the high extinction rates predicted is indirect. Such predictions are therefore to be regarded as hypotheses and it is important to try and establish what are the realities of extinction and the reasons for the apparent survival of so many species in the face of extensive habitat loss, so that the resources of the conservation community can be deployed as effectively as possible to stem the tide of extinctions and minimize losses.

There are many reasons why recorded extinctions do not match the predictions and extrapolations that are frequently published (some of these are considered in other chapters in this volume). These include the different patterns of distribution of species' populations, the different patterns of logging in different parts of the world, the different resilience of species to clearing and fragmentation, their different abilities to occupy secondary habitats, the role of habitat fragments in maintaining species and the effect of corridors to maintain ecolines (*sensu* Bridgewater, 1989), the ability of species to survive long periods (up to a thousand years or more) in small populations, the lack of uniformity of forest vegetation in different parts of the world, and so on. As Thomas (1990) points out, populations that occupy habitat fragments that are far too small to hold thousands of individuals may still possess great conservation potential, particularly when the populations are not completely isolated. He quotes the work of Jones and Diamond (1976) and Soule *et al.* (1988) which suggest that bird populations of above 200 individuals have a high probability of survival for at least 75 years and that some populations can exist at even lower levels for many years.

Known extinction rates are very low. Reasonably good data exist only for mammals and birds, and the current rate of extinction is about one species per year (Reid and Miller, 1989). If other taxa were to exhibit the same liability to extinction as mammals and birds (as some authors suggest, although others would dispute this), then, if the total number of species in the world is, say, 30 million, the annual rate of extinction would be some 2300 species per year. This is a very significant and disturbing number, but it is much less than most estimates given over the last decade.

Recorded extinctions of mammals and birds (Species Database of WCMC) suggest a fourfold increase in the annual extinction rate between 1600 and 1950, but even so the actual numbers were small: 113 species of birds and 83 species of mammal recorded as lost. For vascular plants, the figures, although higher (384), are still low in terms of the total number of 250 000 species known today (Threatened Plants Database of WCMC).

It has to be recognized that these figures all refer to recorded extinctions and are almost certainly much too low, especially for the tropics. There are many reasons for this, including the failure to record systematically all examples reported from the literature (although this would probably not add enormous numbers to the list). There are difficulties too in establishing the facts and an example is given by Woodruff (1991) in a review of Humphrey and Bain's *Endangered Animals of Thailand* (1990) where he regards the information given on the 168 endangered and threatened species as already out of date by the time it was published, due to an increased loss of habitats in the 1980s and during the same period an increase in research by Thai biologists. Woodruff's analysis is more favourable than that of Humphrey and Bain in some cases and less favourable in others: there has been an increase in protected areas from 1.4 to 2.2 million hectares since their report was commissioned. Although the prospects for many Thai species are not assured, the prospects for certain areas and species are stable or are actually improving. This example shows how dynamic the situation can be and how even in areas which have suffered serious depredations, the trends, partly because of increased knowledge and partly through increased environmental protection, have not all been negative. Our lack of detailed knowledge of the distribution of many tropical species has led to their conservation status being revised in the light of further knowledge. While some species have been added to the endangered list, others have been taken off as further populations have been discovered.

We have no knowledge as to the number of species that have become extinct in historical times but which have never been described. Scientific taxonomy only began in the mid-18th century and our knowledge of the tropics in particular was sketchy until this century. Even today many tropical areas are poorly explored and their inventories incomplete or in a few cases non-existent – in some cases, even for relatively well-known groups such as flowering plants and birds. In a recent survey done through the Rapid Assessment Program (RAP) under the auspices of Conservation International (CI) in Bolivia, the Alto Madidi in the Andean foothills was reported to be one of the most diverse rain forests ever studied: for example, 403 birds were recorded including nine species not previously reported for Bolivia and 204 woody plants within a quarter of an acre, many of them new to Bolivia and some new to science. In the Alto Madidi and similar areas we can conclude that the plant, bird and mammal species lists remain very incomplete. Throughout the tropics, many species distributions need to be revised, and statements about endemism and conservation status need constant revision. For some countries, estimates of endemism cannot be made with any degree of accuracy, even for mammals or flowering plants. Clearly, forests are disappearing at too fast a rate for complete assessments of the taxa they contain to be carried out, even for the well-known groups. It is for this reason that rapid surveys are essential if unknown regions of high biodiversity are to be identified and conserved, although it has to be recognized that CI's Rapid Assessment Program is not without its critics.

It is impossible to estimate even approximately how many unrecorded species may have become extinct, although there has been severe habitat loss, including of tropical forests, over the past few centuries, not just in recent decades. Particularly severe habitat alteration started around 1500 when the Europeans first extended their quest for wealth and resources on a large scale to other continents. Later, the introduction of plantation economies in the late 19th and early 20th centuries devastated large areas of tropical lowland forests in Africa, Asia and the Caribbean, yet we have little reported evidence of the loss of species that this presumably caused, even though the faunas and floras of some of these areas were quite well known. One source of evidence is the analysis by Ng and Low (1982) based on data in the *Tree Flora of Malaya* which shows that some plant species were almost certainly lost in the conversion of north-western Peninsular Malaysia into rubber plantations after the classic collections by King were made. As Richards and Tucker (1988) observe, historians of

European colonial systems have often analysed the expansion of agriculture but have rarely paid explicit attention to the other two dimensions of the land: forests and grasslands. Even less attention seems to have been paid to the losses that tropical agriculture based on introduced species entailed.

Moreover, if we assume that today's tropical forests occupy only about 80% of the area they did in the 1830s, it must be assumed that during this contraction, very large numbers of species have been lost in some areas. Yet surprisingly there is no clear-cut evidence for this. One obvious source of evidence is from the taxonomic community who are engaged in producing Floras and Faunas of many parts of the tropics and who are therefore the most likely to become aware of species having become apparently extinct. Despite extensive enquiries we have been unable to obtain conclusive evidence to support the suggestion that massive extinctions have taken place in recent times as Myers and others have suggested. On the contrary, work on projects such as Flora Meso-Americana has, at least in some cases, revealed an increase in abundance in many species (Blackmore, pers. comm. 1991). An exceptional and much quoted situation is described by Gentry (1986) who reports the quite dramatic level of evolution *in situ* in the Centinela ridge in the foothills of the Ecuadorian Andes where he found that at least 38 and probably as many as 90 species (10% of the total flora of the ridge) were endemic to the 'unprepossessing ridge'. However, the last patches of forest were cleared subsequent to his last visit and 'its prospective 90 new species have already passed into botanical history', or so it was assumed. Subsequently, Dodson and Gentry (1991) modified this to say that an undetermined number of species at Centinela are apparently extinct, following brief visits to other areas such as Lita where up to 11 of the species previously considered extinct were refound, and at Poza Honda near La Mana where six were rediscovered. But, as they point out, much more extensive floristic data than are currently available are needed to determine just how many species exist and to what degree they are endemic. Gentry (1986) says that the Centinela ridge does not seem to represent an unusual situation and notes that every isolated cloud forest in Panama has locally endemic species.

However, it would be very rash to attempt to extrapolate from such situations. For example, Dodson and Gentry (1991) suspect that much of the flora of western Ecuador is based on a different set of evolutionary parameters from that of large expanses of relatively homogeneous tropical zone habitats. These include extremely rapid speciation, adaptation to survive with very low population densities, adaptation to small habitat patches, and perhaps rapid natural current extinction rates. This clearly contrasts with the situation in the Brazilian Amazon rain forests, where many species are reportedly widely distributed, so that loss of habitat does not necessarily lead to the loss of a comparable number of species (Bruce Nelson, pers. comm. to VHH).

Another area where a similar situation to that of western Ecuador may obtain is Mount Kinabalu in Sabah, Malaysia (Beaman and Beaman, 1990). Such species with localized distributions are clearly susceptible to extinction if that area of the forest in which they occur is severely disturbed or converted. Peter Ashton (pers. comm. to VHH, 1991) also draws attention to the vulnerability of the coastal lowlands of equatorial regions which often contain islands of endemism, such as in Perak State in north-west Peninsular Malaysia, and the eastern coastal sandy hills and swamps which reach their greatest extent in Johor State. These areas have already undergone such extreme conversion that a number of species discovered by early collectors have almost certainly disappeared.

On the other hand, Ashton and Leong (1990) draw attention to the situation in Malaysia where much of the old lowland dipterocarp forest has been converted into monospecific plantations but notes that Ng and Low's (1982) recent list of potentially endangered species is rather short and gives evidence that much of the tree flora is widespread and constant. Again in Sarawak, recent surveys (US Congress, 1989; ITTO, 1990) suggest that although the commercial timber of the hill forests will have been almost completely harvested by the year 2000 at current rates of harvesting, it appears that much of Sarawak will still be largely covered with forests since the majority of trees are not harvested, the regrowth of pioneer trees is rapid and the canopy is quickly restored. Another report on Sarawak by WWF (1989) states that selective logging has unknown consequences for wildlife populations and may even in some circumstances lead to their increase. Access gained though logging roads has, however, allowed villagers to cultivate forested hill areas that would previously have been inaccessible, and this clearly needs careful management.

We must be cautious, therefore, in our assumptions that loss of habitat necessarily leads to a large-scale loss of species diversity. Yet such assumptions are at the basis of several methods for predicting future patterns of species extinctions, as discussed later.

To bring this point home more clearly, if one recent model to predict extinction rates is correct (Simberloff, 1986), then we are committed to a loss of 12% of the world's plant species (up to 30 000 species) and 15% of the world's bird species (1350) between the mid 1980s and 2015. There is no way of confirming or denying this rate of loss among plants, since we do not have sufficient information on their status and distribution. However, for birds, Simberloff's (1986) model can be tested. Collar and Andrew (1988) produced the first complete and thoroughly researched world list of threatened birds, amounting to a little over 1000 species. Careful scanning of the list indicates that, principally as a result of conservation action, relatively few of these species are likely to become extinct by 2015, though many will be maintained in a precarious position. The annual extinction rate among birds is probably still less than one per year; to corroborate Simberloff's (1986) model, the rate would need to be 45 species per year, which is observably not the case. Of course, it can be argued that the rate of extinction may be low to begin with but later accelerates as populations are reduced to critically low levels. Simberloff's model has more credence if we consider his numbers as referring to species becoming committed to extinction (which is his emphasis in any case), rather than actually dying out during the specified time period or soon thereafter. This is discussed again later.

Much of the evidence for extinctions derives from islands, which is not surprising in view of the vulnerability of island habitats to human intervention, invasion by exotics, extensive agriculture in lowland areas, population pressure (including tourism) and other factors (Bramwell, 1979; Heywood, 1979). Some 40 000 species of higher plant (i.e. a sixth of the world total), for example, are endemic to oceanic islands so that such habitats are a major source of possible extinctions and a considerable percentage of these species are rare or endangered. A similar situation obtains for birds.

Olsen (1989) makes the case for islands having already suffered a catastrophic loss of species through human intervention. He has calculated the number and percentages of extinctions of resident land birds for New Zealand and the Chatham Islands, excluding the species known or likely to have colonized the islands since the arrival of humans. He records 52 resident extant land birds, 12 historical extinctions and 32 prehistoric extinctions. Therefore 46% of the original fauna is now extinct and 33% of the fauna became extinct historically. The prehistoric extinction of 51% of the total number of native land birds of the Hawaiian archipelago has been documented (Olsen, 1989) and in addition 17 species have become extinct in the historic period so that for the total known avifauna, 69% of the species are extinct. In the case of seabirds there are also documented extinctions but most of the reductions have been in numbers and sizes of populations. The prehistoric and historic extinctions of plants are less well documented. Yet, in the Canary Islands, which possess a relict flora, many of the endemic species survive in very small populations and are known to have done so for over a century (Bramwell, 1990). In Mauritius, however, the only hope of the continued survival of many of the 300 plus endemic plant species is intensive management of the pockets of remaining natural vegetation in which they occur, as mini-reserves (Strahm, 1989) supplemented by *ex situ* cultivation and propagation (Jackson *et al.*, 1988).

What can be said about the pattern of documented extinctions since 1600? There are at least three aspects that are not usually mentioned.

1. There have in reality been very few documented extinctions. The total number of species recorded as lost is very small in relation to global biodiversity. This is not to condone what has happened in the past. Every extinction should be considered a tragedy, and the world has lost some remarkable species of plants and animals. What is certain is that very many more species have been seriously depleted in terms of population size and genetic variation.

2. Most documented extinctions are attributable to direct pressures on the species in question, in particular hunting and introduced predators on islands. Relatively few known extinctions have been caused by habitat destruction, and very few by tropical forest destruction. This is not to say that tropical forest clearance has not become a significant threat, but rather that we may have been remarkably lucky up until now, or that large-scale extinctions have gone unnoticed. Ashton (pers. comm. to VHH) believes that for South-east Asia, those forests which received a first cut of selective logging did not lose many plant species but that many species could be lost if a second cut takes place, although this will depend on when the second cut takes place and on what basis. If it is after 30–70 years, as specified in silvicultural rules for dipterocarp forests, then it is likely that many plant and animal species will survive (Johns, Chapter 7; Whitmore, 1990).

3. A large number of the extinctions that are known to have taken place are in the temperate zone. This serves as a warning. Since temperate habitats are frequently the most impacted by human activities (and it should be remembered that massive deforestation first took place in the Mediterranean region several thousand years ago), more tropical extinctions are to be expected in the future if, as seems likely, tropical habitats become modified to the same extent. In view of the rapidly increasing human populations in most tropical countries, this pattern of loss could well be repeated with a much larger number of consequent extinctions, if only because of the greater amount of biodiversity in tropical habitats than in temperate regions. That is not to say, however, that these will reach the scale of some current predictions.

It therefore appears that theory and reality do not converge on the matter of extinction rates. Either the theories are incorrectly formulated, the data are misleading, the time scale wrong, or, quite possibly, a combination of all three.

A focus on species extinction, as we have seen, overlooks another widespread and serious problem – that of population loss and genetic erosion in more widespread species. Genetic impoverishment of species populations, from what has been observed or reported in the literature, must affect a majority of species, so that the 'quasiextinction' rate almost certainly exceeds even the highest published forecasts of species' extinctions. Moreover, the usefulness of using species as a measure of biological diversity at the taxonomic level has been queried by several biologists recently (e.g. Bond, 1989) and the risk of neglecting important population variation has been discussed by Gentry (1990) and Heywood (1991).

COMMITMENT TO EXTINCTION

A critical concept in the matter of predicting extinctions is the notion of species or populations being 'committed to extinction'. This is the phraseology that is used by Simberloff (1986) and followed by Reid and Miller (1989), and implies that when species–area relationships reach an equilibrium, particular species are destined eventually to decline and become extinct either locally or totally. As discussed below, it does not imply any particular time scale for such extinction to take place, nor, for that matter, for an equilibrium to be attained. Simberloff (Chapter 4) notes that there is no accepted theory for the rate at which species will be lost through time and a new equilibrium achieved. In any case, this would be difficult, since susceptibility to extinction varies from species to species so that no generalizations are possible. When Reid (Chapter 3) proposes a revised figure for plant extinctions based on a reworking of the species–area relationship and revised rates of forest loss, his statement is that 'If forest loss continues to accelerate, by the year 2040 some 17–35% of tropical forest species could be committed to eventual extinction when equilibrium numbers are reached'. He is *not* saying that 17–35% of tropical forest species will be extinct by the year 2040 and he draws attention to the various ways in which species loss can be slowed. Commitment to extinction, with all its difficulties, is therefore a different matter from 'imminent extinction' or whatever other phraseology is used which implies that actual loss has or is about to take place any day.

Myers (1986), for example, cites the Pacific Coast forests of coastal Ecuador, the Atlantic forests of Brazil, and Madagascar, in each of which 'as far as we can judge, there were at least 10 000 plant species before broadscale deforestation intervened' and says that 'how many species became extinct as a result of the . . . deforestation is unknown, but must be many thousands, at the very least'. In other formulations Myers states that massive extinctions are about to take place. It is crucial from the point of view of the time scale for conservation action to be clear whether massive extinctions have already happened, are imminent, or both. It makes a very real difference from a practical point of view whether we have a year, ten years or a hundred years in which to take preventive action. For instance, it may well be that deforestation in Ecuador, Brazil and Madagascar has so far caused very few species' extinctions, but that thousands are committed to eventual extinction because of the habitat loss and fragmentation that has already taken place. Dodson and Gentry (1991) comment that the conservation scenario in western Ecuador is grim, but they submit that the conservation of even the tiny fragments that remain there might well conserve a substantial proportion of the estimated original complement of 1260 plant species endemic to the area. However, they warn that such action must be taken very soon, for no matter how adept these species may be in surviving in tiny fragments of vegetation, all will surely be lost if these remnants of forest go.

There is considerable evidence that the notion of 'commitment to extinction' has validity and requires further investigation. As mentioned above, Simberloff (1986) estimated, using a species–area model, that some 1350 species of birds would be committed to extinction by the year 2015. To carry out a very preliminary test of this prediction, we undertook a brief analysis of the world list of threatened bird species (Collar and Andrew, 1988). The list includes 1029 species, of which 12 were almost certainly extinct prior to the list being completed (but were included because in each case there is a very remote chance of rediscovery). In looking through the list, we attempted to identify those species, which, on current trends are likely to be 'committed to extinction' by the year 2015. We interpreted 'committed to extinction' as referring to any species whose populations in the wild are no longer viable and which will inevitably become extinct, unless there is major conservation action to reverse the trend (through techniques such as habitat restoration, re-introduction, elimination of introduced predators, captive breeding, etc.). Species on the world list of threatened species that are not, in our view, yet committed to extinction are those which can survive in viable populations providing their habitats are conserved and hunting and collecting pressures are controlled. Our estimate (admittedly imperfect) is that 450 bird species will be committed to extinction by 2015, and that 27 of these could be extinct already (this does not include the 12 almost certain extinctions referred to above). This number is 33% of that given by Simberloff; however, this discrepancy is much smaller than those noted earlier in this chapter. It should be noted that many of the 450 species are not associated with tropical forest. Also, habitat loss is only one of the causes of decline; factors such as hunting, collecting, disease and introduced species have very major effects as well. Species are particularly at risk when several of these factors are combined.

One reason for the discrepancy between our estimates and those of

Simberloff is that we could have underestimated the increasing pressures that species will face over the next 20–30 years. Another reason could simply be the problems in using species–area curves to predict species extinctions in the first place, as outlined by Simberloff (Chapter 4) as well as later in this chapter. However, the important point that results from our brief analysis is that it is clear that the number of bird species being added to the critically endangered list is increasing rapidly. Although actual extinctions remain low, the number of species that are going to require very major investments if they are to be saved is becoming very large. If action on behalf of these species is not stepped up, and very soon, then we are likely to see a considerable increase in the actual extinction rate. It should be stressed that genuine and successful recovery programmes for endangered birds (or for any other critically threatened species for that matter) have been very few. Most conservation programmes for such species have had the effect of preventing further decline, but not actually bringing about recovery; the species remains as imperilled as ever.

Another example of species lingering much longer than would normally be predicted relates to plants in the Mediterranean region. In the Mediterranean basin, which has about 25 000 plant species, about a quarter of which are endemic (Quezel, 1985), despite deforestation, land clearance, grazing, intensive agriculture and urban and industrial development, some of which date back thousands of years, few species are recorded as having been lost in historical times although large numbers are rare or endangered, often existing in small populations on mountain tops (Gómez-Campo, 1985). Many of the surviving species might well be committed to extinction, according to our definition. As Greuter (1991) aptly comments, 'Many endangered species appear to have either an almost miraculous capacity for survival, or a guardian angel is watching over their destiny! This means that it is not too late to attempt to protect the Mediterranean flora as a whole, while still identifying appropriate priorities with regard to the goals and means of conservation.'

It is clear, from the data on birds, that the process of extinction can be a lengthy one for any given species. In fact this is probably the norm, especially for higher plants. We already have a situation in which habitats have become so fragmented that many species and populations are now isolated from each other. If these populations are small, then genetic erosion will inevitably take place in most cases. Frequently, such loss of genetic variation can itself fuel the extinction process, through loss of vigour, and loss of the capacity to adapt to environmental changes.

It is beyond doubt that deforestation has led, and continues to lead, to major losses of genetic variability within species populations and there is field evidence to indicate that many species (just how many is not known) are reduced to populations that are below any of the suggested minimum viable levels for long-term survival. The loss of biodiversity at the infraspecies level may not be so easy a concept to put over to administrators and politicians as loss of species (although who is going to provide the list of even a thousand recent species' extinctions when asked for?) but it is now widely accepted as important for crop, forestry and livestock species. For species of economic importance in which genetic variation is essential for breeding and selection, genetic erosion is a serious issue on the world stage with major implications for the future security of our food supply. When one considers that about 5000 plant species are cultivated by humankind and at least twice as many more are used from the wild, it is clear that loss of genetic variation in such a large number of species alone is a cause for considerable concern and demands action. To ignore it in other species not known at present to have direct human value would be scientifically irresponsible.

If we are able to understand better the ways in which species and their populations respond to different kinds and patterns of habitat loss in different kinds of forest and in different regions of the world, their ability to survive in small numbers and maintain allelic variation in fragmented habitats or in secondary communities, their capacity to migrate and maintain gene flow between distant sites, and the site-specific key interactions involved in the reproduction and regeneration of forest species, we stand a much greater chance of understanding the nature and time scale of the extinction process. This information should assist us in taking effective conservation action to secure the survival of as many species as possible in the long-term future. We also need to study the ways in which conservation action has affected species extinctions so that our predictions for the future can be more soundly based. Some of these aspects are discussed below.

PREDICTING SPECIES LOSS

Most estimates of species loss have restricted the focus to tropical forests, since this is where the great majority of species is believed to occur. While this is probably a perfectly reasonable approach, it often becomes overlooked in subsequent discussion. It also ignores the fact that the majority of documented extinctions are from outside the rain forest biome. All previous estimates of the global extinction rates have been derived from estimates of rain forest loss. Myers (1990) has, however, extended his hot-spots analysis of 10 tropical forest areas (Myers, 1988b) to cover an additional four tropical forest areas and four Mediterranean-type zones and concludes that the 18 areas together, support about 50 000 endemic plant species. Insofar as these areas are threatened with imminent destruction of their natural vegetation, Myers states that we could witness the elimination of at least 25 000 plant species (or 10% of the earth's total) in these areas alone unless increased conservation measures are implemented forthwith.

We accept for the purposes of this chapter the recent estimates of forest loss published by WRI (1990) and reproduced with modifications by Reid (Chapter 3). However, it must be stressed that the empirical relationship between deforestation and species loss is not known and has not been established experimentally (Lugo, 1988). There are, in fact, numerous intractable difficulties in predicting extinction rates.

First, we need to consider the use of species–area curves to predict species extinctions. Reid and Miller (1989) have reviewed in some detail the methods of estimating species extinction rates. The difficulties of applying the species–area relationship to predicting extinction rates are discussed by Simberloff (Chapter 4) who notes that it is not so much a single theory as an empirical generalization, although he regards it as a 'worthwhile exercise . . . to attempt a first guess at how many extinctions deforestation will generate in tropical forest'. There is considerable debate in the literature on the equilibrium theory of island biogeography and its applicability to conservation (Simberloff, 1988 for a useful review). Boecklen and Bell (1987) consider that it is a doubtful representation of natural processes that are often complex and species specific. This theory is central to most extinction rate models. However, Saunders *et al.* (1991) point out that the species–area equation is of little value in managing fragmented ecosystems. For example, the equation may give a manager a rough idea of how many species will be maintained on a remnant in a given area 'but will yield absolutely no information on the practical issue of which habitats contribute most to species richness or on which species are most likely to be lost from the remnant.' Our concern in this chapter is with gaining a sufficient understanding of extinction dynamics to guide management for conservation in practical ways. As Simberloff (1986) states, it is 'sad that unwarranted focus on island biogeography theory has detracted from the main task of refuge planners, determining what habitats are important and how to maintain them.'

Methods using a species–area curve lead to a rule of thumb calculation that a loss of 90% of a habitat leads to a 50% loss of species. There remains considerable debate concerning the nature of species–area curves, including the different equations that can be used to describe curves in different locations (Simberloff, Chapter 4 and Reid, Chapter 3). It could be that these differences are merely artefacts of random processes (possibly resulting from the canonical lognormal distribution of species abundances). If this is the case, then it could be highly speculative to extrapolate downwards or upwards from such curves.

Moreover, the species–area curve (in a mainland situation) is nothing more than a self-evident fact: that as one enlarges an area, it comes eventually to encompass the geographical ranges of more species. The danger comes when this is extrapolated backwards, and it is assumed that by reducing the size of a forest, it will lose species according to the same gradient. One reason why this might not be the case is that species are not distributed at random. The key factor is often the loss of forest in specific, identifiable sites which are of particular importance for species diversity and endemism. If any of these sites are completely cleared, then species loss will be much higher than the rate predicted by the curve. For example, the conversion of the tiny Sinharaja Forest in Sri Lanka would lead to the loss of 60% of the endemic flora of the country. Conversely, if such centres of diversity and endemism are protected, then species loss can be considerably reduced. It also needs to be emphasized that species–area curves provide no insight into rates of extinction, nor indeed into the identities of the species most likely to be at risk. Yet, from a conservation viewpoint, this information is particularly important.

Quite apart from the difficulties associated with models to predict species extinctions, there are a number of other inherent difficulties associated with the prediction of extinction rates, as follows:

1. We have nothing more than approximate indications as to the total number of species currently extant, even for some relatively well-known groups. In particular it is almost impossible to estimate rates of loss for very poorly known groups, such as some invertebrates and soil fungi. It should be noted, however, that both the arguments and the evidence for extinctions are based primarily on selected groups such as vertebrates, birds and vascular plants. The rarity and extinction of some species may in fact be an artefact of taxonomy. An acutely difficult problem is highlighted by Gentry (1986) who suggests that speciation leading to the high degree of endemism found in many tropical forests, such as those of the Ecuadorian Andes referred to earlier, arises through different causes from that found in the Mediterranean/temperate regions such as California. He notes that 'conservationists may be faced with the philosophical question of whether preservation of such randomly generated and often rather nondescript species should be accorded the same priority as preservation of a *Sequoia*.' He later (1990) raised the question of comparability of species-level taxa from different parts of the world and argued for a narrow species concept to ensure that recognizably different populations should be given attention through their recognition as species. It is quite clear that we need to interpret species numbers (and their possible extinction) with a great deal of caution when making global estimates or regional comparisons.
2. We have no accurate estimate of the number of species that occur in

rain forests, still less how these are distributed in various types of forest, although we would not dispute the overall figures given by Raven (1987) and Myers (1988a, 1990) for higher plants in certain areas. But, as noted above, geographical distributions are very incompletely known in the tropics, the sizes of floras and faunas are very uncertain, and degrees of endemism are only approximations, thus making calculations and predictions difficult. The very localized distributions of many tropical forest species may in some cases be an accident of sampling.

3. We do not know the validity of taxonomic extrapolations. Most estimates are based to some extent on mammal and bird data, and yet we cannot say whether it is valid to extrapolate these to plants and invertebrates. It could be that mammals and birds are more vulnerable to extinction, since larger species occurring at lower densities are more prone to decline to levels below a minimum viable population size, especially in fragmented habitats. Conversely, insects, with their narrow ecological specializations, might be more at risk from human disturbance of their habitats.

4. We do not know the validity of geographic extrapolations. Most studies of the effects of tropical forest clearance and fragmentation on species loss have been carried out in the Neotropics. One might expect Neotropical forests, with their often greater species richness in many cases than, say, African and some Asian forests, to be more prone to species loss (though diversity and stability theories might indicate the reverse). It should also be noted that tropical forests have been subject to very different kinds of exploitation and management in different parts of the world.

5. The population size of most tropical species is unknown. There is no substitute for a thorough knowledge of species population sizes if probabilities of extinction are to be estimated. Many species exist as small populations today of 500, 100 or even fewer individuals, either through natural causes or as a result of habitat loss and other threats. Moran and Hopper (1987) suggest that in the ancient landscape of south-west Western Australia it may be that species have evolved to cope with small population sizes. Many small populations of localized species of *Eucalyptus* in this region show considerable genetic diversity and Moran and Hopper suggest that the oft-quoted minimum population size of 500 seems inappropriate for these species. The small populations of localized species now restricted to natural remnants of vegetation can, they suggest, be conserved in these reserves for over 1000 years and maintain a high percentage of their genetic variation. We do not know how many such species there are, although globally, in other taxa that occur naturally in such small populations, it may well run into thousands.

6. We know very little about the adaptability of primary forest species to secondary forest. We also know very little about the long-term effects of habitat fragmentation on extinction. If the case of the localized species of *Eucalyptus* (Moran and Hopper, 1987) and the high amount of genetic variation found in their small populations is found to be of more general applicability, then remnants of natural vegetation may play a more significant role in the conservation of genetic resources than has been appreciated in the past. The problem is that each individual species is likely to respond differently to any such changes, and usually in an unpredictable way. For most tropical forest species we know too little of their autecology to be sure of their survival requirements.

7. We have no global understanding of the impact that conservation measures have had on extinction rates, nor how effective they might be in stemming further extinctions. This is considered later in this chapter.

8. We are usually unable to confirm that extinction has taken place until long after the event. There are many records of species which have been presumed extinct for 50 years or more being rediscovered. For example, in Malaysia the fern *Ptericanthes grandis*, discovered in 1898, was rediscovered in 1987 (Latiff, pers. comm. to VHH, 1989). There are many avian examples: for instance, the yellow-throated serin *Serinus flavigula*, last recorded in Ethiopia in 1892, was rediscovered in 1989. However, the converse can also be true, and species thought to be safe can decline without anyone noticing. The extinction process can extend over a very long period and as Western (1989) aptly comments, 'enormous crippling damage to species and ecosystems can happen long before the finale'.

9. We do not know to what extent extinctions continue to 'work though' in an area, long after deforestation has been stopped. There is experimental evidence to demonstrate that such 'relaxation' occurs in species communities following forest fragmentation. However, we do not know how large forest patches need to be to avoid this phenomenon, nor how long the 'relaxation' takes to 'work through'. This relates to the issue of 'commitment to extinction' which was discussed earlier.

The foregoing discussion should suffice to demonstrate that the prediction of extinction rates is an almost impossible task. Many previous attempts to estimate global extinction rates have involved a 'guesstimate' at the total number of species in the world, an implied assumption that birds and mammals are in general representative of all species (inherently unlikely), an assumption that species communities in tropical forests are similar the world over, a failure to address the issue of species adaptability to secondary habitats, no mention of the effects of conservation action, and the use of species–area curves for purposes for which they were never intended. Because of these and other reasons outlined above, we feel that we cannot attach any great degree of confidence to any predictions of species extinction rates. Field evidence suggests that most estimates might be too high. But the field evidence might be misleading and the estimates could be too low. We simply have no way of knowing. We are a very long way off having any clearer idea and our methods of collecting and synthesizing information are quite inadequate.

Fundamental to this question of global extinction rates is the attempt to come up with a single global statistic to describe an extraordinarily multivariate situation. There are so few data points available that theory is likely to remain way out ahead of the facts for some time. In any case, it is likely that the path to predicting extinction rates will not be through global modelling, but through acquiring and synthesizing better distributional and autecological knowledge of threatened taxa, systematic surveys of particular key families or other taxonomic groupings, and an understanding of fragmentation of habitats on survival of species populations.

Be this as it may, we believe that too much focus on global extinction rates, current or future, risks diverting attention from the more immediate issues such as the large numbers of species and populations that are reduced in size and variability as a result of habitat loss or fragmentation and are therefore in need of urgent conservation action *now*. Other important issues we should address are: what have been the effects of conservation and management measures in alleviating species loss; and what additional measures are required to reduce the rate to as low a level as is feasible.

EFFECTS OF CONSERVATION ACTION

All the evidence suggests that the loss or conversion of tropical forests will continue at a high rate despite initiatives such as the Tropical Forestry Action Plan and other efforts to slow it. We should not, however, underestimate the effectiveness of well-directed conservation action on the survival of many species, albeit in smaller populations than we would wish, and we should take vigorous action now to retain as many options as possible for the future. The establishment of protected areas in sites of high biodiversity remains perhaps the most critical factor in slowing rates of extinction.

One example comes from the Cape Floristic Province of South Africa. This is an area of exceptional richness – 8600 plant species in the fynbos, over 73% (6300 species) of which are endemic and 17% rare or threatened. In a review of the preservation of species in southern African nature reserves, Siegfried (1989) found that an extraordinarily high percentage of southern Africa's terrestrial vertebrates and about 74% of its vascular plants are represented by breeding populations in the region's nature reserves, including the fynbos, even though they were not expressly designed to conserve biological diversity. These reserves must surely reduce or at least slow the rate of expected extinctions over the coming decades. It is worth mentioning that one of the most serious threats to the remaining fynbos comes from invasion by exotic trees such as pines and hakeas, which adds another dimension to the management problems involved in conserving such areas.

In most studies of extinction rates, the effects of conservation action have not received much attention. The impression given has always been that conservation activities are so small scale that their results are insignificant in relation to the global problem. While it is undoubtedly true that conservation measures have had only a minor impact on global deforestation rates, it does not necessarily follow that their positive effects in reducing species extinction rates are minor. This is because, as mentioned above, species distributions in tropical forests are far from random. There are a number of critical sites for species diversity and endemism, and if these can be protected, large portions of the global biodiversity will be secured. This is the underlying principle of IUCN's Centres of Plant Diversity Project.

This is exactly what has happened as a result of conservation action. Some 5–6% of the tropical forest biome is now protected in reserves. This percentage is rising rapidly and has its distribution heavily skewed towards the most diverse sites (at least as indicated by bird and mammal distributions). Conservationists have deliberately pressed for the conservation of sites of high biodiversity, and as such have played tricks with the species–area curve. Because of the careful positioning of many protected areas, the actual rate of species extinction has probably been considerably reduced. This is not cause for complacency. The protected area system in tropical forests is still far from complete, and there are some particular areas that give cause for concern: Philippines, Burma, Ecuador, Guyana, Bolivia, Guinea, Liberia, Congo, Madagascar and the three Indochinese countries are examples of countries with very incomplete and poorly enforced protected area systems in which extinction risks must be high.

It is still too early to answer quantitatively the question 'what have been the effects of conservation measures in alleviating extinction rates?' A preliminary study in Africa suggests that over 95% of the Afrotropical avifauna occurs within the continent's protected areas. This still needs verification, but for birds and mammals, it ought to be possible to assess the effectiveness of the global protected area network in protecting them.

If such a study is carried out, it should then be possible to identify the gaps that need to be filled with additional reserves. An urgent requirement is the preparation of an inventory of the biodiversity in protected areas.

A second phase of this could aim to assess the effectiveness of the protected area system in conserving viable populations of species and 'minimum dynamic areas'. The mere presence of a species in a protected area does not mean that its population there is at a viable level, and indeed many species of plants and animals occur in much smaller populations than any of the recommended minimum viable populations. However, a number of actions can be taken to deal with inviable populations, providing the problem is identified well in advance, and we do have a few good examples of successful management of such populations. Up until now, however, this has not generally been the case. The 450 bird species mentioned earlier that are likely to be committed to extinction by 2015 are all species that are already, or will soon be, surviving only in inviable populations. It is with these species that the conservation movement has had the least success, because in most cases the traditional protected area approach is no longer sufficient to ensure their survival. More drastic forms of management intervention are required on their behalf, as mentioned earlier. However, the protected area approach remains the best way to conserve most species, and is probably the only way to prevent even more species declining to levels that require very heavy investments to ensure their survival.

In general, the effects of conservation action in reducing species loss have probably been underestimated. This might be because since conservation action has so far not been conspicuously successful in broader objectives (such as integrating conservation and development needs, regulating human population growth, slowing down climate change, saving water catchments, controlling soil erosion, etc.), it is assumed that attempts at controlling species loss have been similarly wanting. However, there is still no room for complacency, and this has principally to do with the problem of population fragmentation.

MANAGING FRAGMENTED POPULATIONS

In the past, the tropical forest biome has covered enormous areas of the tropical zone. As long as such large areas of habitat survived, the future of the species they contained was relatively secure. However, it now seems incontestible that by the middle of the next century, the remaining undisturbed forest will be reduced to a number of smaller patches, which have been established as protected areas. While this is obviously better than nothing, there will inevitably be severe management problems associated with the inherent attributes of small populations of animals and plants. Particular problems are presented by small, fragmented populations, and these can be summarized as follows: genetic; demographic; ecological; and accidental.

The *genetic* problems associated with small populations are becoming well known. Small populations of species restricted to a tiny fraction of their former range may represent only a small amount of the original genetic variation within the species, although, as noted above, this need not necessarily be so. In addition, if these populations are below a minimum viable population size, then they may be subject to inbreeding, and the continued loss of genetic variation through drift, rather than natural selection. Inbred populations may lose vigour, in terms of reproductive fitness and resistance to disease. Moreover, such populations lose the capacity to evolve rapidly to new situations, and this is particularly serious at a time when global climate change will probably require species to make rapid adjustments. The genetic erosion of species is therefore likely to become an increasing problem. It can be dealt with in a number of ways. Obviously, it is easiest if protected areas are made as large as possible, since this will minimize the need for expensive management actions later on. On the other hand, Gentry suggests that in the Amazon, local endemism of plants and animals is the result of edaphic specializations with fragmentation during arid periods of the Pleistocene. This has the conservation implication that rather than going for the conservation of large refugial areas so as to conserve many unrelated taxa, it would be better to focus on areas of high habitat diversity. Where large areas cannot be conserved, the management of fragments is an option that will need increasingly to be considered, as Dodson and Gentry (1991) suggest for western Ecuador. Other options include establishing corridors (Saunders and Hobbs, 1991), maintaining artificial gene flow between fragmented populations, and establishing *ex situ* populations to reinforce wild stocks.

The *demographic* problems are particularly acute in very small populations. For animal populations of less than 20 individuals, the risk of having all or most of the young in one particular generation of the same sex is greatly increased. This in itself relates to the genetic problems, since if one individual makes a disproportionate contribution to the next generation, then a genetic bottleneck occurs, further contributing to the genetic erosion in the population. To solve the demographic problems, the same actions are needed as for the genetic problems.

The *ecological* problems can be very complex and difficult to detect, especially in tropical forest communities. Frequently, a forest patch does not contain all the resources required for a species throughout its annual cycle. Many tropical forest animal species move quite widely over the year due to the patchy distribution of food supply. If these movements are prevented by forest fragmentation, local extirpation may follow. This is believed to be one of the principal causes of 'relaxation' in fragmented species communities. Similar problems occur with plant species and their dependence on pollinators and dispersers. Other ecological problems associated with fragmentation include increased predation rates. Many non-forest predators range some distance into forests, and their impact is clearly much more severe, the smaller the protected area is. There is often very little that can be done about some of these ecological problems. Certain resources can be added to a small reserve quite easily (such as salt-licks). But large fruiting trees cannot be moved. Again, the solution depends on areas being large enough in the first place for the whole system to be self-supporting. One practical management option is to maintain habitat corridors between the reserve and important resources that are missing or are in short supply in the reserve, or to establish a matrix of reserves.

The *accidental* problems are the most difficult to control. Once the remaining area of habitat is small, the chances of something unexpected going wrong are enormously increased. A hurricane hitting Puerto Rico is unlikely to cause any species extinctions if the forest cover is 80%. But if only 1% remains, the whole area can be devastated. The same applies to disease, fire, poaching outbreaks, clearance of the area due to a breakdown in law and order, and increasing human population pressures on the area. Again, the only solution is to make the protected area large enough in the first instance to minimize the chances of accidents, or to maintain as complete a mosaic of fragments as possible, if the former option is no longer available.

Behind each of the four problems outlined above is the need to have protected areas as large as is humanly feasible, though we recognize that it is already too late in some areas to exercise this choice. Although the establishment of protected areas in critical sites is a key first step to minimizing species extinction, we can still expect more problems in the future, especially if remaining reserves become isolated in a 'sea' of inhospitable agricultural habitat.

In the long term, the following three steps will be needed to avert mass extinctions.

1. Identify areas of high concentrations of biodiversity.
2. Complete the establishment of a global protected area system that includes the above, to include viable populations of as many species as possible.
3. Prevent the protected area system from becoming too fragmented, in particular by maintaining areas of sustainably managed secondary forest through as much of the existing tropical forest as possible.

This third step might prove difficult to achieve in view of rapidly expanding human populations. However, if large areas of tropical forest do not survive, then severe perturbations in the world climate can be expected, that will have a heavy human and conservation cost. The ultimate disaster would be if the habitats in tropical forest protected areas degenerate before our eyes for climatic reasons. Then we really shall see a massive number of extinctions. Maintaining as much as possible of the existing forest cover is essential if such a disaster is to be averted.

CONCLUSIONS

We draw the following main conclusions from the foregoing discussion:

1. It remains very difficult to define the relationship between habitat area and species survival, and therefore to estimate extinction rates. Hence, all attempts to calculate such rates, and the resulting figures, are highly uncertain.
2. Perhaps our most important conclusion is that the fragmentation of habitats, and hence also of species populations, is resulting in the loss of genetic variation. We believe that this affects an even larger number of species than those that have been estimated to be at risk of extinction. We contend that, in consequence, the size and seriousness of the problem confronting us is much greater than has previously been assumed. We emphasize the importance of taking genetic conservation fully into account in the planning of national conservation programmes.
3. Conservation organizations like IUCN and WWF need to face up to the challenges posed by habitat fragmentation, and design, promote and implement a massive programme of conservation action, both inside and outside protected areas, involving *ex situ* methods as appropriate.
4. In the long term, the most effective means of preventing tropical forest extinctions will be to complete the establishment of the protected area network in centres of diversity and endemism, and to maintain as much as possible of the remaining forest area under sustainable forest management. The risk of accelerating climatic changes means that unsustainable forest management practices have to be curtailed as a global priority. The obstacles to progress in this area are rooted in social, developmental and economic factors that need to be addressed as a matter of urgency.
5. The emphasis of future research should be placed on assessing the short-term and long-term effectiveness of conservation actions in reducing the risk of extinction of the increasing number of severely genetically depleted populations of species. Such research should aim to have practical outputs, enabling managers to plan ahead for conservation measures that will be needed to maintain small, fragmented species

populations. As Saunders *et al.* (1991) suggest, we should encourage research and discussion focusing on practical issues relating to the impact of fragmentation on natural ecosystems and managing remnants for conservation.

We believe that there is still time for practical steps to be taken that will stem the loss of species, although sadly we cannot envisage a scenario in which extinctions can be totally halted. However, if appropriate policies are adopted now, it should be possible to reduce greatly the biological impoverishment of the World.

SUMMARY

In this chapter we review the current evidence for mass extinctions, principally as a result of tropical forest loss. We conclude that there is currently little evidence of extinctions at the rates predicted by some theoretical models. However, we consider that there is ample evidence that the number of species that will require urgent action to prevent their eventual disappearance is increasingly rapidly. These species are described as 'committed to extinction', and these will require major interventions if they are to be saved. We also draw attention to the increasing numbers of species with fragmented populations and which are almost certainly suffering severe genetic erosion. We consider that these trends could be setting the stage for much higher rates of extinction in the future.

ACKNOWLEDGEMENTS

Many colleagues throughout the world have contributed information and advice during the preparation of this chapter. The help of the following is especially acknowledged: Peter Ashton, Stephen Blackmore, Cal Dodson, Al Gentry, Arturo Gómez-Pompa, Martin Holdgate, Jeff McNeely, George Rabb, Peter Raven, Sy Sohmer and Tim Whitmore. The views expounded in the chapter are, however, our own and do not necessarily reflect the opinions of those mentioned above, unless directly cited in the text.

REFERENCES

Ashton, P. and Leong, Yueh Kwong (1990) Box 1. Conservation in Malaysia. *Trends in Ecology and Evolution*, 5, 395.

Beaman, J.H. and Beaman, R.S. (1990) Diversity and distribution patterns in the flora of Mount Kinabalu, in *The Plant Diversity of Malesia* (eds P. Baas, K. Kalkman and R. Geesink), Kluwer Academic Publishers, Dordrecht, pp. 147–60.

Boecklen, W.J. and Bell, G.W. (1987) Consequences of faunal collapse and genetic drift to the design of nature reserves, in *Nature Conservation: the Role of Remnants of Native Vegetation* (eds D.A. Saunders, G.W. Arnold, A.A. Burbidge and A.J.M. Hopkins), Surrey Beatty Pty Limited, Chipping Norton, NSW, pp. 141–9.

Bond, W.J. (1989) Describing and conserving biotic diversity, in *Biotic Diversity in Southern Africa, Concepts and Conservation*, (ed. B.J. Huntley), Oxford University Press, Cape Town, pp. 2–18.

Bramwell, D. (ed.) (1979) *Plants and Islands*. Academic Press, London.

Bramwell, D. (1990) Conserving biodiversity in the Canary Islands. *Ann. Missouri Bot. Gard.*, **77**, 28–37.

Bridgewater, P.B. (1989) Connectivity: an Australian perspective, in *Nature Conservation: the role of remnants* (eds. D.A. Saunders, G.W. Arnold, A.A. Burbidge and A.J.M. Hopkins), Surrey Beatty Pty Limited, Chipping Norton, New South Wales.

Collar, N.J. and Andrew, P. (1988) *Birds to watch. The ICBP checklist of threatened bird species*. ICBP Technical Publication No. 8, Cambridge, UK.

Dodson, C.H. and Gentry, A.H. (1991) Biological extinction in western Ecuador. *Ann. Missouri Bot. Gard.*, **78**, 273–95.

Ehrlich, P. (1988) The loss of diversity. Causes and consequences, in *Biodiversity* (eds E.C. Wilson and F.M. Peter), National Academy Press, Washington, DC, pp. 21–7.

Gentry, A.H. (1986) Endemism in tropical versus temperate plant communities, in *Conservation Biology* (ed. M.E. Soule), Sinauer Associates, Sunderland, MA, pp. 153–81.

Gentry, A.H. (1990) Herbarium taxonomy versus field knowledge. Is there an attainable solution? *Flora Malesiana Bulletin Special*, **1**, 31–5.

Ginzberg, L.R. (1990) Reconstructibility of density dependence and the conservative assessment of extinction risks. *Conservation Biology*, **4**, 63–70.

Ginzberg, L.R., Slobodkin, L.B., Johnson, K. and Bindman, A.G. (1982) Quasiextinction possibilities as a measure of impact on population growth. *Risk Analysis*, **2**, 171–81.

Gómez-Campo, C. (1985) The conservation of Mediterranean plants: principles and problems, in *Plant Conservation in the Mediterranean Area* (ed. C. Gómez-Campo), W. Junk, Dordrecht, pp. 3–8.

Greuter, W. (1991) The need to preserve genetic resources, in *The Conservation of Wild Relatives of Cultivated Plants*. *Environmental Encounters* Series, No. 8, Council of Europe, Strasbourg.

Heywood, V.H. (1979) The future of island floras, in *Plants and Islands* (ed. D. Bramwell), Academic Press, London, pp. 431–41.

Heywood, V.H. (1991) Needs for stability of nomenclature in conservation, in *Improving the Stability of Names: Needs and Options* (ed. D. Hawksworth), Regnum Vegetabile No. 123, pp. 53–8.

Humphrey, S.R. and Bain, J.R. (1990) *Endangered Animals of Thailand*. Sandhill Crane Press, Gainesville.

ITTO (1990) *The Promotion of Sustainable Forest Management: a Case Study in Sarawak, Malaysia*. International Tropical Timber Organization, Yokohama.

Jackson, P.S., Wyse, Strahm, W., Cronk, Q.C.B. and Parnell, J.A.N. (1988) The propagation of endangered plants in Mauritius. *Moorea*, 7, 35–45.

Jones, H.L. and Diamond, J.M. (1976) Short-time-base studies of turnover in breeding bird populations in the Californian Channel Islands. *Condor*, 78, 526–49.

Lovejoy, T. (1989) Editorial. The obligations of a biologist. *Conservation Biology*, 3, 329–30.

Lugo, A. (1988) Diversity of Tropical Species. Questions that elude answers. *Biology International Special issue* 19.

Moran, G.F. and Hopper, S.D. (1987) Geographic population structure of *eucalyptus* and the conservation of their genetic resources, in *Nature Conservation: the Role of Remnants* (eds D.A. Saunders, G.W. Arnold, A.A. Burbidge and A.J.M. Hopkins), Surrey Beatty Pty Limited, Chipping Norton, NSW, pp. 151–67.

Myers, N. (1986) Tackling mass extinction of species: a great creative challenge. XXVI. The Horace M. Albright Lectureship in Conservation. University of California, College of Natural Resources Department of Forestry and Resource Management, Berkeley.

Myers, N. (1988a) Tropical forests and their species. Going, going . . .? in *Biodiversity* (eds E.O. Wilson and F.M. Peter), National Academy Press, Washington DC, pp. 28–35.

Myers, N. (1988b) Threatened biotas: 'hot-spots' in tropical forests. *Environmentalist*, 8, 187–208.

Myers, N. (1990) The biological challenge: extended hot-spots analysis. *Environmentalist*, 10, 243–56.

Ng, F.S.P. and Low, C.M. (1982) Check list of endemic trees of the Malay Peninsula. *FRI Kepong Research Pamphlet* no. 88.

Olsen, S.L. (1989) Extinction in islands: man as catastrophe, in *Conservation for the Twenty-first Century* (eds. D. Western and M. Pearl), Oxford University Press, New York, pp. 50–53.

Quezel, P. (1985) Definition of the Mediterranean area and the origin of its flora, in *Plant Conservation in the Mediterranean Area* (ed. C. Gómez-Campo), W. Junk, Dordrecht, pp. 9–24.

Raven, P.H. (1987) The scope of the plant conservation problem world-wide, in *Botanic Gardens and the World Conservation Strategy* (eds D. Bramwell, O. Hamann, V. Heywood and H. Synge), Academic Press, London, pp. 10–19.

Raven, P.H. (1990) The politics of preserving biodiversity. *BioScience*, 40, 769.

Reid, W.R. and Miller, K.R. (1989) *Keeping Options Alive. The Scientific Basis for Conserving Biodiversity*. World Resources Institute, Washington.

Richards, J.F. and Tucker, R.P. (1988) Introduction, in *World Deforestation in the Twentieth Centurey* (eds. J.F. Richards and R.P. Tucker), Duke University Press, Durham and London, pp. 1–12.

Saunders, D.A. and Hobbs, R.J. (eds.) (1991) *Nature Conservation 2: the Role of Corridors*, Surrey Beatty and Sons, Chipping Norton, New South Wales.

Saunders, D.A., Hobbs, R.J. and Margules, C.R. (1991) Biological consequences of ecosystem fragmentation: a review. *Biological Conservation*, 5, 18–32.

Saunders, D.A., Arnold, G.W., Burbidge, A.A. and Hopkins, A.J.M. (eds) (1990) *Nature Conservation: the Role of Remnants*. Surrey Beatty Pty Limited, Chipping Norton, New South Wales.

Siegfried, W.R. (1989) Preservation of species in Southern African nature reserves, in *Biotic Diversity in Southern Africa: Concepts and Conservation* (ed. B. Huntley), Oxford University Press, Cape Town, pp. 186–201.

Simberloff, D. (1986) Are we on the verge of a mass extinction in tropical rain forests? in *Dynamics of Extinction* (ed. D.K. Elliott) Wiley, New York, pp. 165–80.

Simberloff, D. (1988) The contribution of population and community biology to conservation science. *Ann. Rev. Ecol. Syst.*, **19**, 473–512.

Soule, M.E., Bolger, D.T., Alberts, A.C., Wright, J., Sorice, M. and Hill, S. (1988) Reconstructed dynamics of rapid extinction of chaparral-requiring birds in urban habitat islands. *Conservation Biology*, **2**, 75–92.

Soule, M. (ed.) (1987) *Viable Populations for Biology*. Cambridge University Press, Cambridge.

Strahm, W. (1989) *Plant Red Data Book for Rodrigues*. Koeltz Scientific Books, Konigstein for IUCN.

Thomas, C.D. (1990) What do real population dynamics tell us about minimum viable population sizes? *Conservation Biology*, **4**, 324–7.

Tree Flora of Malaya (1972) Volumes 1 and 2 (ed. T.C. Whitmore), Longman, Kuala Lumpur and London.

Tree Flora of Malaya (1978) Volume 3 (ed. F.S.P. Ng), Longman, Kuala Lumpur and London.

US Congress (1989) *The Tropical Timber Industry in Sarawak, Malaysia*. Congressional Staff Study Mission to Malaysia. Superintendent of Documents, Congressional Sales Office, US Printing Office, Washington DC.

Western, D. (1989) Conservation Biology, in *Conservation for the Twentieth Century* (eds D. Western and M. Pearl), Oxford University Press, New York and Oxford, pp. 31–6.

Whitmore, T.C. (1990) *Introduction to Tropical Rain Forests*. Clarendon Press, Oxford.

Wilson, E.C. (1988) The current state of biological diversity, in *Biodiversity* (eds E.O. Wilson and F.M. Peter) National Academy Press, Washington, DC, pp. 3–18.

Woodruff, D.S. (1991) Thai Red Data Book [Review of Endangered Animals of Thailand by S.R. Humphrey and J.R. Bain], *Conservation Biology*, **5**, 131–2.

WRI (1990) World Resources 1990–91. A report by the World Resources Institute in collaboration with the United Nations Environment Programme and the United Nations Development Programme. Oxford University Press, New York and Oxford.

WWF (1989) *Rainforest Conservation in Sarawak: An International Policy for WWF*, (eds M. Kavanagh, Rahmin Abdullah Abdul and C.J. Hails). WWF Malaysia, Kuala Lumpur.

6

Habitat alteration and species loss in Brazilian forests

K.S. BROWN Jr. and G.G. BROWN

TROPICAL BIODIVERSITY, EVOLUTIONARY EROSION AND SPECIES LOSS

The extinction of species, especially vertebrate animals and economic plants, has dominated discussion in conservation circles for over a century. There is no doubt that post-Pleistocene humankind has eliminated many species from those habitats occupied and transformed into human-directed channels for energy, food, shelter and other resources. Especially in the case of endemics of islands, watercourses and lakes, and of large predators, highly specialized or hunted species, entire and irreplaceable biological units, including species and even orders, have disappeared forever from natural systems under direct or indirect human pressure.

Recently, attention has been focused on the seemingly inevitable reduction of the high biological diversity of tropical forests, as a result of their occupation, use and destruction by rapidly growing human populations. Widely publicized models and predictions have led to a considerable body of mythology, exaggeration and misinformation, spread around the world by headline-hungry media, often with little factual basis in reliable observations and controlled experimentation. Indeed, the few measurements of real species losses in fragmented tropical forest systems are strongly discordant with this body of popular and attractive myth, as will be described.

On the other hand, some authors have noted that even local population extinctions above the natural turnover rate, reduce the biodiversity of the community, change the course of evolution, and not infrequently lead to permanent restructuring of the entire ecosystem, with ongoing 'cascade' effects and appreciable influence on neighbouring communities. Genetic diversity, the base for evolution and adaptation, is very sensitive to minor disturbance in the fabric of the system; and as this diversity is highly concentrated in small invertebrates, especially arthropods which occupy easily modified microhabitats in the soil, vegetation and canopy, it may be eroded locally at a rate far higher than that observed in the regional extinction of larger animal or plant species.

Conversely, the actions of humankind can also increase genetic diversity, both at the local population and regional species levels, through creation of secondary successions, new selective pressures, resource alteration, microhabitat multiplication, and niche diversification. Furthermore, in some ecosystems, including islands and seasonal tropical forests, strictly natural phenomena can lead to massive genetic erosion and species extinction. Thus, although appearing to be deceptively easy, it may in fact be extremely difficult to measure the overall impact of humankind on biodiversity, evolutionary potential, and species turnover, except in the obvious cases of total habitat elimination or small, isolated systems.

This chapter will focus on recent data on habitat alteration in the two Brazilian tropical rain forest biomes, Amazonian and Atlantic, and the effects of this alteration on local and regional biodiversity, especially of animals, including indicator insects (Brown, 1991). An attempt will be made to test and evaluate a number of models and predictions, and to isolate several factors relating habitat alteration to changes in biodiversity, addressing the following questions:

1. how far has destruction by human populations of these two Brazilian tropical forest biomes progressed? How much species and evolutionary erosion can be seen as a result of this progressive destruction?
2. what are the effects of regional history and disturbance level on the response of biodiversity to habitat alteration?
3. does the island-biogeography theory help to predict loss of species diversity in reduced or fragmented continental systems of tropical forest?
4. how do topography and land-use methods affect the relationship between habitat alteration and biodiversity?
5. how does the reduction of biodiversity affect the human populations associated with tropical forest systems?
6. what can be learned from these patterns, to help plan for future human

---

[a] Average rate of deforestation from 1978 to 1989 was 21 218 ± 10% km²/year, peaking in 1987 and dropping appreciably since then (Fearnside *et al.*, 1990). Preliminary data for 1990 indicate only a little over 10 000 km² was cleared.
[b] Approximate total area within the forested part of each State considered here (see also Figs 6.1 and 6.4) of parks, reserves, ecological stations and heritage areas.
[c] Includes, in Pará and Maranhão, 97 643 km² of older deforestation, mostly in the Belém-Bragança area from the last century, much now in high secondary forest.
Sources: Fearnside *et al.* (1990); Tardin *et al.* (1990); SOS Mata Atlântica/INPE/IBAMA (1990).

Table 6.1. Deforestation (in km^2) in the Brazilian Amazonian and Atlantic regions

(a) Amazonian region

State (abbrev.) (NW-NE-SW-SE)	Total area in 'Legal Amazonia'	Approximate original area of forest	Total area deforested by August 1989[a]	Percentage of forest still intact	Totally protected areas 1990[b]
Amazonas (AM)	1 567 954	1 540 000	21 551	99	70 000
Roraima (RR)	225 017	185 000	3 621	98	9 000
Amapá (AP)	142 359	112 000	1 016	99	10 860
Pará (PA)	1 246 833	1 140 000	139 604[c]	88	17 300
Maranhão (MA)	260 233	163 000	88 664[c]	46	3 410
Acre (AC)	153 698	153 000	8 836	94	6 820
Rondônia (RO)	238 379	214 000	31 476	85	17 380
Mato Grosso (MT)	802 403	417 000	79 594	81	2 620
Tocantins (TO)	269 911	40 000	22 327	44	6 000
Amazonian Totals	4 906 787	3 964 000	396 689[a,c]	90	143 390

(b) Atlantic region

State (abbrev.) (NE-SW-SE-S)	Total area	Approximate original area of forest	Total area deforested by 1990	Percentage of forest still intact	Totally protected areas 1990[b]
Rio Grande do Norte (RN)	53 167	2 320	1 986	14	20
Paraíba (PB)	53 958	6 960	6 528	6	80
Pernambuco (PE)	101 023	18 500	18 106	2	60
Alagoas (AL)	29 107	14 200	13 880	2	190
Sergipe (SE)	21 862	9 800	9 758	1	30
Bahia (BA)	566 979	114 800	107 896	6	1 390
Espírito Santo (ES)	45 733	45 500	40 919	10	910
Minas Gerais (MG)	586 624	288 000	274 700	5	4 740
Goiás (GO)	340 166	38 200	35 858	6	–
Mato Grosso do Sul (MS)	357 472	61 700	50 884	16	–
Rio de Janeiro (RJ)	43 653	43 500	38 499	11	4 420
São Paulo (SP)	248 256	208 000	175 790	15	11 540
Paraná (PR)	199 324	183 000	148 664	19	5 160
Santa Catarina (SC)	95 318	82 400	58 670	29	1 690
Rio Grande do Sul (RS)	280 674	88 900	77 618	13	720
Atlantic Totals	3 023 316	1 205 780	1 059 756	12	30 950

use of tropical forests, minimizing the loss of biodiversity and evolutionary potential?

7. is sustained harvesting of tropical forest resources a realistic goal, permitting sustainable development of the associated human populations?

DEFORESTATION IN THE BRAZILIAN AMAZON AND ATLANTIC REGIONS, 1500–1990

Table 6.1 and Figs 6.1 and 6.2 present the most reliable and recent information available on the extent of the two major blocks of Brazilian forest, before European colonization, and today after nearly 500 years of intensive occupation by an increasingly dense and industrialized human society. Much of the conversion of natural habitat to anthropic systems was complete by the end of the 19th century in north-eastern and south-eastern Brazil (including the Belém–Bragança region east of the mouth of the Amazon), while the central Atlantic region (southern Bahia and northern Espírito Santo) and the still very limited amounts of the Amazon forest have mostly been converted through the opening of new access routes in the past 25 years. For comparison, there is also included (Fig. 6.3) the probable maximum extent of disintegration of the same forests through natural climatic change at the peak of the most recent long cold, dry glacial episode (Würm–Wisconsin), 18 000 years ago, in accord with a model which uses palaeoclimatic, geomorphological, soil and vegetational evidence to deduce probability of forest persistence under unfavourable climatic regimes interpreted with the aid of present-day observations where these regimes prevail in central and northern Brazil (Brown, 1979, 1982). This last Pleistocene fragmentation episode has been variously interpreted by different authors as leading to extensive species and population extinctions, extensive species multiplication, or to simple differentiation at various infraspecific levels. A fragile consensus, and a certain correlation with biogeographical patterns, suggests that all these processes, with an emphasis on the last, took place, as they do today in dynamic natural systems, though perhaps somewhat intensified in certain regions and taxonomic groups (Turner, 1982; Brown, 1982, 1987; Endler, 1982).

The Brazilian Amazon forests are still little affected by outside occupation, with only about 10% of their area converted to anthropic systems (Figs. 6.1 and 6.2, Table 6.1), in spite of wildly exaggerated reports and predictions of their imminent disappearance in the press, and even in reputable scientific journals and books. One quarter of the area is already under effective conservation regimes (Fig 6.4) and it is expected that up to 70% will be maintained in natural forest for climate and water protection. In 1990, 0.25% of the area was destroyed in mostly marginal,

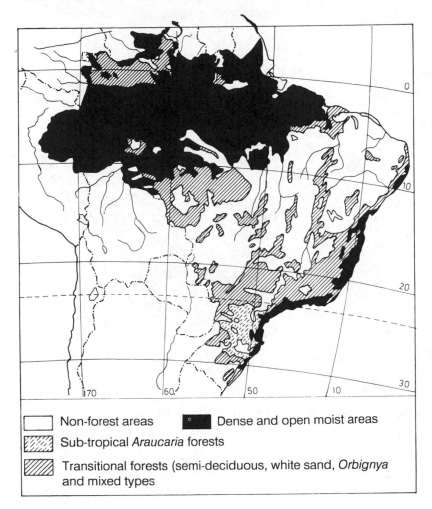

Figure 6.1. Extent of the Brazilian Amazon and Atlantic tropical moist forests before European colonization (1500 AD).

drier or more seasonal forests, poor in endemic species, an amount which has been lower every year since 1987. Nevertheless, 10 000 km^2 (or one million hectares) of conversion per year represents ample occasion for local elimination of species, disruption of systems, and erosion of biosynthetic and evolutionary capacities of the biota. Several projects in various parts of the Amazon are studying the dynamics of disturbed or fragmented forest systems (Lovejoy *et al.*, 1986), necessarily on a very small scale (less than 10 000 ha). These have shown, predictably, that large or specialized, higher-trophic-level species are most sensitive, but

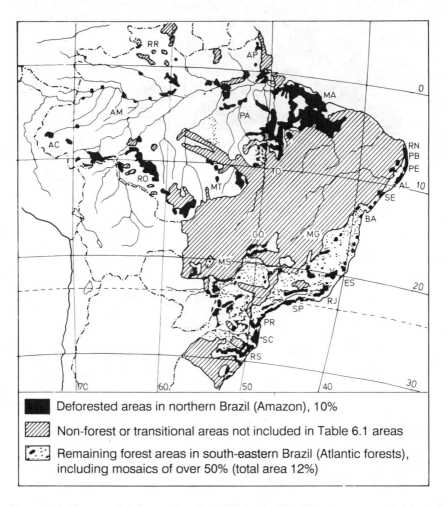

Figure 6.2. Extent of deforestation by 1990 in the Brazilian Amazon and Atlantic tropical moist forests. State names abbreviated to two letters (see Table 6.1).

these are usually also the most widespread and mobile, and able to emigrate or persist in nearby intact systems. Weedy species are encouraged and may increase after human disturbance or fragmentation of the forest. Appreciable changes also occur in the hydrology, soils, and vegetation structure of smaller fragments and adjacent areas, indicating that the local system will not soon return to its former state. The disturbed areas are often even richer in understorey species, which occur in denser populations, but deep-forest species may be permanently lost from the site or region. In the lowland Amazon, most supposedly 'point endemic'

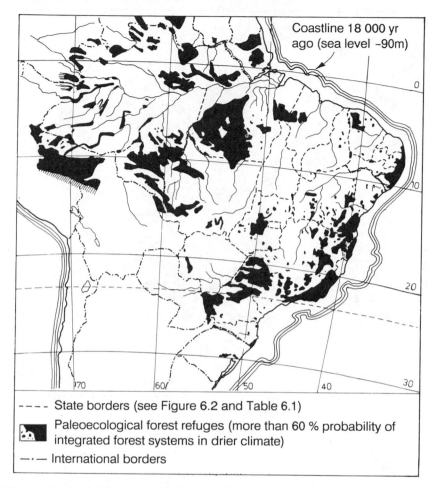

Coastline 18 000 yr ago (sea level –90m)

- - - - State borders (see Figure 6.2 and Table 6.1)

Paleoecological forest refuges (more than 60 % probability of integrated forest systems in drier climate)

—·— International borders

Figure 6.3. Probable extent of the Brazilian Amazon and Atlantic forests about 18 000 years ago, according to a model derived from geoscientific analyses (soils, geomorphology, paleoclimate, vegetation structure). (From Brown, 1979, 1982, 1991.)

species have been shown to be widespread though local, but in montane or special habitats, such as open-vegetation enclaves, local extinction of some species might occur, in addition to the local adaptive genes of more widespread species.

Because so little of the Amazon has been explored biologically, and most of the species and subspecies occupy ranges of more than 10 000 km^2, it is nearly impossible to evaluate the effects of the still very limited deforestation on loss of species. The inevitable local loss of genetic

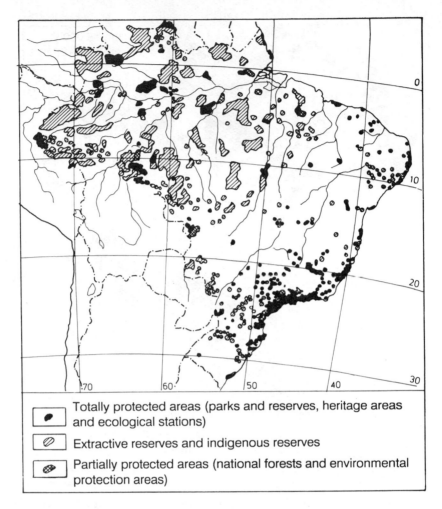

Figure 6.4. Conservation units in Brazil in 1990. (From Brown, 1991, and many other sources). National forests established in the Amazon region in 1989–90 (84 800 km²) are not included.

diversity is undoubtedly greater than under natural disturbance regimes, but an increase in diversity is also produced by the new selective regimes of disturbed, microheterogeneous habitats. It seems futile to try to relate habitat alteration in the Brazilian Amazon to eventual species loss, under the present circumstances and with the data in hand.

The more complex Atlantic coastal systems, mostly with higher species diversity than most of the Amazon forests, have been reduced to about 12% of their original extent (Figs. 6.1 and 6.2, Table 6.1). A widely publicized model for the state of São Paulo (Victor *et al.*, 1975) suggests that only 3% of the original vegetation is likely to remain in the year 2000. Far more broken up than at the height of the last glaciation (Figs 6.2 and 6.3), the Atlantic forests should represent an ideal system to measure the effects of deforestation on species extinction. Island biogeographical theory would predict a loss of about 50% of the species, perhaps somewhat more or less because of the strongly fragmented character of the remaining natural vegetation. It is widely proclaimed that all species endemic to the Atlantic forests (average near 50% overall endemism, as high as 95% in some groups in this very ancient, reasonably isolated system) should be regarded as threatened. What can be seen in the persistence or loss of species in this system?

SPECIES LOSSES IN THE ATLANTIC FORESTS

In 1989, members of the Brazilian Society of Zoology with long experience in the field prepared an updated list of Brazilian animals threatened with extinction. This list (Bernardes *et al.*, 1990), officially published in December 1989, includes all animals believed to be in the three most threatened IUCN categories (vulnerable, endangered and extinct): 24 primates (and one additional subspecies), 13 carnivores, four edentates, two manatees, two whales and a dolphin, six mice and a hedgehog, three deer, 107 birds (and an additional subspecies), six turtles, one snake, two caymans, a frog, 23 butterflies (and three additional subspecies), four damselflies, a *Peripatus*, and a coral. Of these 202 species, all but 31 (half of these primates) inhabit the Atlantic forests or offshore islands. Only six of these 171 Atlantic species were listed as possibly extinct, namely two birds (a parrot unofficially observed in recent years and a very inconspicuous becard probably overlooked in most surveys as the voice is not known), and four butterflies (placed formally in this category because not seen in over 50 years by the very few observers available, but still with many suitable habitats intact and unexplored). Thus, the group of zoologists could not find a single known animal species which could be properly declared as extinct, in spite of the massive reduction in area and fragmentation of their habitats in the past decades and centuries of intensive human activity. A second list of over 120 lesser-known animal species, some of which may later be included as threatened, show no species considered extinct; and the older Brazilian list of threatened plants, presently under revision, also indicated no species as extinct (Cavalcanti, 1981).

One possible reaction to this declared persistence of the 50% of the Atlantic species which ought to be extinct, is to ascribe it to ignorance or lack of data on the part of the biologists. Few invertebrate groups or members of the soil biota are well enough known to permit adequate definition of threatened species. Bats, fishes and most anurans were omitted. The updated and detailed list of threatened plants has not been completed and older collections are fragmentary, with many species only known from their type collection, which was often from a restricted habitat later destroyed. Perhaps many naturally rare or restricted species, never described, are already gone; but these are not subject to any verification.

Closer examination of the existing data on both well- and little-known groups, however, supports the affirmation that little or no species extinction has yet occurred (though some may be in very fragile persistence) in the Atlantic forests. Indeed, an appreciable number of species considered extinct 20 years ago, including several birds and six butterflies, have been rediscovered more recently, often far from their typical localities, but always in the same habitat type, which was specifically sought out and explored in order to check for remnant populations of these little-known animals and plants (Brown, 1991).

The most likely reasons for the persistence of these rare species in the fragmented Atlantic forest landscape, which have important implications for conservation planning and habitat management in this region and other similar biomes around the world, are several.

1. The highly complex topography of the Atlantic region, breaking the biome up into many different small microhabitats, repeated across the landscape in similar situations of soil, exposure and drainage, macroclimate and disturbance, and giving a true surface area far larger than the cartographic (flat view) area. This means that most Atlantic species have always lived in small but viable populations in small habitat fragments, presumably moving between them through other unfavourable or resource-poor habitats, and adapted to this style of life.
2. The high degree of natural disturbance in these habitats, on short and long time scales leading to adaptive responses in many populations. These disturbances result from the rolling topography and heavy rains, varying seasonality and cold spells, macroclimatic fluctuations, and unpredictable dry, hot or wet periods, associated with a complex interaction between pressure cells, winds, and orographic features.
3. The appreciable plasticity of most plants and animals, especially those with short generation times and abundant gametes and zygotes, permitting rapid adaptive shifts and efficient tracking of environmental changes.

In this way, it can be seen that certain geological and biological factors greatly influence the effects of habitat alteration on species loss. Once again, it must be emphasized that destruction of 90% of a biome, even if leading to minimum extinction of entire species, must result in the elimination of many local populations, with the inevitable erosion of genetic diversity implied by this. Even this erosion can be greatly reduced, however, by a proper application of landscape ecology, involving social and economic land-use factors interacting with innate geological and biological ones.

THE INFLUENCE OF DIFFERENT PATTERNS OF HUMAN LAND USE ON SPECIES LOSS

The fundamental biological values of subspecific endemism (regional differentiation) and species diversity are distributed in a non-correlated manner in Brazilian forests (Fig. 6.5; Brown, 1982). The former, well correlated with the geoscientific model for forest reduction in the late Pleistocene (Fig. 6.3) is concentrated rationally into a number of well-defined centres (Brown, 1979, 1982) or 'environmental conformities' (Endler, 1982), where species diversity is only moderate. At the species level, endemism, at least in the insects analysed, is clearly associated with only four barely overlapping regions in all the neotropics, two of these corresponding to the two major blocks of tropical moist forest in Brazil (Brown, 1982, 1991). Species diversity peaks in scattered points of high environmental heterogeneity, usually on the peripheries of the endemic centres (Fig. 6.5). In these peripheral regions are usually found the 'paleoenvironments' which concentrate very ancient and rare species in many animal and plant groups (Brown, 1991), including a disproportionate number of those considered as threatened with extinction.

From this pattern, it follows that effective biological conservation will require that large areas be carefully surveyed, managed and monitored, to include all three types of biological parameters (endemism, diversity and rarity) in their non-coincident localities. The values for these may vary dramatically even on a very small scale and be quite unpredictable without actual measurement in the field on a good day in the right season. It thus becomes especially important to maintain as much of the landscape as possible under potential conservation regimes, at least in a 'patchwork environment' (Norton, 1985), until the biotic resources and their distribution are known and understood.

Conservation in such complex, microheterogeneous systems is thus a delicate exercise in landscape ecology, where human populations and natural environments interact in a mosaic of low-intensity sustainable resource exploitation, without excessive habitat conversion or channelling of productivity. Such usage occurs in most indigenous and extractive reserves in the Amazon (Fearnside, 1989a; Schwartz, 1989; Alegretti, 1990; Almeida, 1990), and also in small-farm mosaics in the states of Santa Catarina, Espírito Santo, and other regions in the Atlantic forests. It

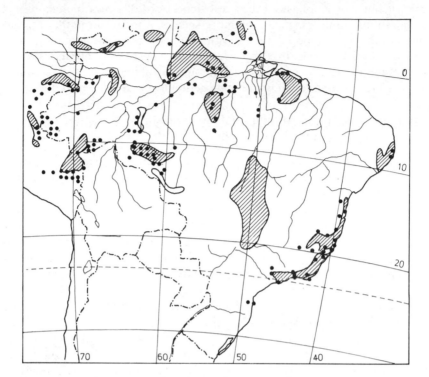

Figure 6.5. Brazilian areas of highest subspecies endemism (hatched areas have two-thirds of the maximum corrected endemism value for each endemic centre), and points of highest species diversity (black dots), using over 100 species of aposematic butterflies in three ecologically different forest groups, and nearly 1000 localities grouped into quadrants of 30′ of longitude and latitude (Brown, 1982, 1991).

definitely does not occur, and indeed genetic erosion becomes critically intense, when large blocks of land – especially in regions of heterogeneous topography or vegetation – are treated in a more uniform fashion, such as in large Amazonian cattle ranches with their orthogonal, manicured edges enclosing thousands of hectares of pasture (Fearnside, 1989b), or in the plantation-economy monocultures of the Atlantic region – sugar cane in the nearly totally converted northeast (Table 6.1) or the interior of São Paulo, coffee in the latter and in the Minas Gerais/Espírito Santo hill country, bananas along the coastal plain, reforestation with exotic economic trees in many regions, and even cacao in southern Bahia, which leaves part of the forest canopy intact but completely transforms the lower storeys.

These types of·land use, so hostile to rational or sustainable exploitation of natural resources and therefore constantly expanding to destroy more systems and area, usually represent large-scale political or economic interests linked to 'development' schemes (Norton, 1985), with the immediate benefits generally accruing to a few investors at the top of a denigrating social system (Suzuki, 1990). They result in a maximum of genetic erosion, even when combined with enforced 'conservation' of part of the area, which is inevitably strongly degraded by hunting and logging, often to increase further the immediate material return.

In contrast, a model of low-intensity mosaic usage of the landscape by extractive or indigenous populations results in a minimum of genetic erosion and maximum of conservation (Schwartz, 1989; Alegretti, 1990). Here the density is usually less than one person per km^2 but it may eventually attain up to ten times this in carefully planned or managed mosaics, thus already similar to small-landowner agriculture. Enhancement of biological values is even possible in strongly heterogeneous systems where a good part of the managed productivity is channelled back to nature (Brown and Cardoso, 1989 and unpublished results). Such 'undeveloped' land and resource use, often decried as 'primitive, uneconomical and predatory' by the official 'development' agencies (the true predators), has been shown in fact to be the most economically profitable use of certain Amazon forests in the medium to long run (Peters *et al.*, 1989, Table 6.2). Although it does not serve the interests of denser urban populations or powerful (but usually shortsighted) economic groups (Fearnside 1989b), it does maintain biodiversity (Fig. 6.6) and natural processes effectively and, if kept in harmony with an industrialized sector, permits an adequate living standard for a reasonable number of (often unwitting) mini-conservationists in the rural scene. Such landscape-use patterns must be encouraged, or even required, in tropical forests, along with sufficient transportation of the sustainably extracted products to suitable markets, to minimize the impact of human occupation and habitat alteration on natural resources which may be needed in the future.

THE ROOTS OF DEFORESTATION AND OTHER IRRATIONAL, ANTIECONOMICAL LAND-USE PATTERNS

Fearnside (1987) has given extensive lists of the proximal and underlying causes of deforestation in the Amazon, namely land speculation, tax incentives and penalties, other subsidies and loans, exportable production, shifting cultivation and colonist attrition. All these are fuelled by population growth, low land prices, inflation, displacement of small landowners, road-building, politics, fear of the forest, status, immediatism, externalization of environmental costs and depletion, low labour requirements for pasture, and low agricultural yields. All of these causes relate best to the least economical, most destructive, and commonest form of

Table 6.2. Analysis of economic return from different uses of 1 ha of lowland tropical forest in eastern Peru showing the advantages of resources other than wood

Type of use	Yearly income (after labour and transportation)	Net present value (discounted 20-yr)	Sum	Authors' comments
Extractive uses (people-centred economy which maintains or improves the forest)				
Harvest of latex (rubber)	$22	$440 (7%)		Does not include other products, or tourism
Harvest of edible fruit	$400	$6000 (93%)		
			$6440	
		(leave 25% in forest for regeneration)		
'Sustainable' selective logging	$15	$490		Small value
			$6930	
'Development' uses (top-down economy which destroys the forest)				
One-time removal of marketable timber		$1000 (no 20-yr discount)		Cutting destroys extractive resources
Reforestation with *Gmelina arborea*	$159	$3184		Gives much less than extraction; not sustainable
			$4184	
Intensive cattle ranching on ideal pasture	$148	< $2960		Gives much less than extraction: not sustainable
				maintenance costs not included

Source: Peters *et al.*, (1989).

deforestation in Brazil, namely conversion into pasture (Fearnside, 1987), a practice which must be outlawed as soon as possible, since all its effects are noxious, even to those who may seem to obtain an immediate gain. Suzuki (1990) has condensed these many causes into three emotive words:

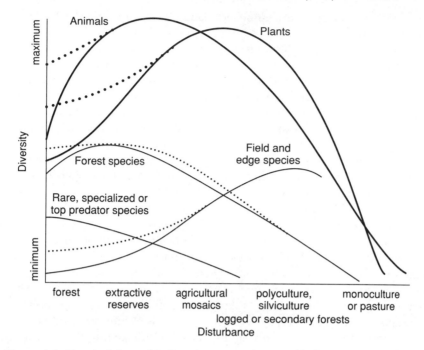

Figure 6.6. Tropical forest biodiversity versus degree of habitat alteration in more homogeneous or low-productivity, and (dotted lines) more complex or high-productivity systems in Brazil.

ignorance, injustice and greed. The draft World Conservation Strategy for the 1990s (Prescott-Allen, 1989) suggests further factors, namely high population growth and increasing and mismanaged environmental demand (p. 8), or simply violence, oppression and want (p. 49). Large economic and development interests, fed by political corruption, generate and thrive under these conditions throughout tropical forests, in a characteristic 'top-down' approach to resource use and mobilization (Thrupp, 1989). Under this model, conservation of genetic resources becomes a low-priority lottery in which all victories are temporary and all defeats permanent. Even those projects like selective logging which claim to seek sustainable use of the resources are usually farcical in the Amazon (Fearnside, 1989b; Keto *et al.*, 1990), since in the Amazon even the most elementary monitoring of these schemes shows the same irreversible system degradation, energy channelling, soil and biodiversity depletion, and bleak future as the conversion into pasture achieves somewhat more directly (Peters *et al.*, 1989; Table 6.2). No matter whether on low-productivity or high-productivity soils, in the Amazon or the Atlantic region, these large-scale projects dovetail into a single desolate picture (Suzuki, 1990) in which human and natural resources are channelled, squandered and destroyed, gravely affecting all populations involved in the process. Indeed, such schemes can only persist in a thoroughly corrupted economic and social fabric.

Fortunately, these homogeneous methods of land abuse are becoming progressively less convincing to governments, scientific agencies and local people, who have learned to work together in some areas to break this pattern of resource alienation, landscape deterioration and future despair.

Outstanding examples of socioeconomic structures leading to resources stewardship for the future can be found in both Brazilian moist forest areas. In Santa Catarina, most exceptional among Brazilian states in having only 5% of its people in the capital city (other states have 15–80% concentrated there), the well-educated and stable population is spread out across the entire landscape in small, intensively managed land areas. This population, occupying both tropical and subtropical forests with high rainfall, has one of the highest rural standards of living in South America, working hard enough to avoid poverty but not wishing to become ostentatiously wealthy or to concentrate economic power, planning for their children and later descendants. About 30% of the state's forests have been maintained (Table 6.1) in large and small patches that preserve essentially the entire biota, recognized as both economically and socially important to the people. Starting in the mid-19th century, as many as 40 000 families in Santa Catarina practised in the recent past a low-intensity farming of insects, supplemented by sustainable extraction, especially of butterflies, used to make tasteful, widely distributed and appreciated art objects. Many other ways of democratic and sustained use of natural resources have helped to maintain an economy and society admired by all Brazilians, avoiding the excesses of strongly unequal distribution of land, money and power which are so inimical to conservation and human values.

The Amazon forests, less complex topographically but more differentiated genetically (Fig. 6.5), need special care and protection against large-scale, ecologically uninformed timber and deforestation projects. In the far western state of Acre, historically isolated rubber-tapping and Brazil-nut-harvesting populations have learned by necessity to live in equilibrium with their forest resources, and today struggle to maintain this low-intensity economic system in as much of the state as possible, through local people's associations, co-operating with other indigenous or riverine forest dwellers, with scientists, and with government, in an efficiently balanced system which has successfully resisted 'imposed development' and deforestation in many areas (Alegretti, 1990). Admitting that people on any level can be corrupted by the empty dreams of material prosperity – governors, scientists, and even forest dwellers have all fallen at times (Redford, 1989) – the three 'powers' work together in a reasonably secure scheme of mutual support and vigilance, to maintain extractive reserves (Fearnside, 1989a) as realistic models of provenly sustainable forest resource use. The economy can grow through scientific research and discovery of new resources with proven markets (oil plants, medicinal herbs, natural fragrances and pigments) which can be used sustainably, but neither this ideal socioeconomic structure nor the genetic resources on which it depends (and which it supports) can survive large-scale 'development' projects in the region. These, and sometimes other less materialistic interests often subvert the integrity of the local community by introducing modern hard technologies, industrialized goods, consumerism, and markets for secondary or non-sustainable forest products, besides directly destroying the forest system. Such tendencies must be watched and carefully controlled.

CONCLUSIONS ON OPTIMAL STRATEGIES FOR CONSERVATION OF SPECIES AND GENES IN TROPICAL FORESTS

It is evident from the foregoing that realistic strategies for conservation of tropical forests and their genetic heritage must pass as much through the economic and social considerations which govern land-use patterns, as through biological survey and monitoring, or description of soil, climate and topographic features. Governments, NGOs and local leaders can recognize certain landscapes, places or biotas that are especially important to be preserved, can even understand, as do indigenous groups, the importance of preserving natural processes rather than products as essential resources (Alcorn, 1989), and can set up adequate incentives, penalties and monitoring to encourage conservation. However, effective use and saving of the resources rests in ultimate analysis with the local people who depend upon them for their present life and future hopes (Table 6.3; McNeely *et al.*, 1990, pp. 47–62). It is tragic that so many politicians, economists and international agencies regard these people as living in 'unsustainable and predatory extractivist' or 'poor rural' economies, and seem to want to disintegrate their systems of resource protection, mobilizing the 'liberated' resources unsustainably and destructively, for the immediate benefits of larger parasitic urban populations or specialized economic interests, often aimed at the export of superfluous items, little related to human adaptation and strongly prejudicial to local adapted societies. Must all the world become like Liverpool, Tokyo or São Paulo? Or will the powerful societies in these cities be able to respect their own future options by maintaining complex mosaics of natural vegetation and low-intensity resource utilization in much of the vast area which they parasitize for their food, clean air and water, research projects and spiritual re-creation?

Table 6.3. Conservation processes (reading left to right) in a 'bottom-up', people-centred economy

| | | Contributions from: | | |
Step	Local people	Scientists using theory	Scientists applying theory	Government action
Initiation of process	Recognition of the necessity for preservation of specific resources	Choose indicator groups, develop biological indices	Classify and evaluate local resources, in preliminary field survey	Should not interfere at this stage, but support local people's groups and research
Designation of best areas for priority in preservation	From local knowledge, indication of concentration of desirable resources; trace history of land use, define priorities	Recognize historical and ecological determinants, apply theory of patch dynamics	Realize detailed surveys of resources and indicators, geological and biological indices	Recognize value of a conservation unit in the region, support process, finance detailed surveys
Propose area(s) for permanent preservation or management regimes	Support for creation of an optimum preserved area, accept responsibility for control over and protection of it	Determine methods for optimization of preservation of populations and processes	Correlate all the theoretical and survey data to suggest the optimum area(s)	Officially recognize the process and formalize the priorities and proposal for a preserved area
Define and describe the conservation unit(s) in region	Encourage wide recognition of the unit(s) in society, effect respect for and protection of preservation	Develop indicators for the dynamic processes and changes	Follow populations and processes throughout the preserved area	Recognize unit and make it official at the higher organization levels

Table 6.3. Continued

Step	Local people	Contributions from: Scientists using theory	Scientists applying theory	Government action
Develop infrastructure and management plans for preserved area(s)	Collaborate with scientists and visitors in increasing values and management of unit(s)	Mobilize theories of population biology, community ecology, temporal dynamics of systems	Define acceptable subareas of the unit for diverse human activities and research	Sanction the official management plan for the area
Continually monitor the conservation unit(s)	Help in resource, population and system monitoring within and around the unit	Apply biogeographical theories, sampling protocols, indicators	Apply and test monitoring methods for optimal information and efficiency; verify island effects	Promote regional and national plans for conservation, management, and monitoring

Source: adapted from Brown (1991) and many other sources.

The earlier cultural subspecies of *Homo sapiens* (Table 6.4), who had to learn to live in relative harmony with their natural resources, still exist in some parts of the world, along with their more derived, sophisticated, presumptuous and much less well-adapted descendants. The latter populations, far from equilibrium with their resources or environments and in their rush towards their own suicidal extinction or at least catastrophic regulation, are forcing the earlier kinds of human society to extinction, along with other representatives of the evolutionary vigour of life on Earth. By respecting the sensitivity to natural diversity and processes inherent to these less sophisticated socioeconomic systems, urban populations may develop a new recognition of these well-springs of their survival, on which they continue to depend for all fundamental resources. If natural diversity is disrespected and destroyed, the demise of the intimately associated human populations will be immediately followed by that of the more distant but still codependent industrial societies. Rubber-tappers, indigenous peoples, small farmers and fishermen, and many other sustainable local communities, know the tropical forest systems and can show the way to their rational use and conservation (Table 6.3; Clad,

Table 6.4. 'Cultural subspecies' of the *Homo sapiens* line

Cultural subspecies	Typical activities	Started (yr BP)	Present percentage of population	Adaptation
H. sapiens aborigine (= *H. habilis, H. erectus*)	Subsistence hunting and gathering, tools, fire	3 000 000	<< 0.1	
migratorius	Same, with extensive migrations; more tools, language, tradition	400 000	<< 0.1	Maximum integration in natural systems, adaptation to nature
include neanderthalensis				
religiosus	Same, with art, ritual, and reflection	80 000	< 0.5	
agropastoril	Domestication of plants and animals, stable artificial environments and controlled resources	20 000	5	Beginning of disintegration of adaptivity
commercialis	Formation of city-states with trade, laws, complex structuring of society	8 000	15	Increase of artificial environments, middlemen, classes, wars
industrialis	Expansion of capital and heavy industry; extensive urbanization	300	20	
energeticus	Abundant use of cheap energy, rapid population growth	100	50	Progressive disassociation from nature
informaticus	Global society in communication on all levels	40	< 10	Modify nature, adapt to cities

Table 6.4. Continued

Cultural subspecies		Typical activities	Started (yr BP)	Present percentage of population	Adaptation
possible future subspecies	*communicator?*	Increase of integration	Today	Not definable	Definitive disassociation from, or effective reintegration with nature
	innovatus?	Channel of aggression to creativity			
	destructor?	Increase of aggression and disintegration			
	durabilis?	Resistant to pressure from environment			
	sentimentalis?	Takes refuge in past simplicity and values			

Source: Adapted from Eccles (1979, pp. 79–122); Boughey (1975); and many other recent sources.

1985), if the urban scientists and governments will so permit. This requires empowering these local communities (Thrupp, 1989) to enable them to continue controlling their resources (Schwartz, 1989) remaining neither in 'enforced primitivism' (Goodland, 1985) nor 'imposed development' (Colchester, 1989), but simply benefiting from the technical advances of the industrialized society without allowing themselves and their invaluable knowledge to be swamped. Those of us who live and work in the technological sectors must be sensitive to the importance of these older societies, and willing to assist in their continual survival for the future of all.

SUMMARY

The relation between habitat alteration and loss of biodiversity is examined in the Amazon and Atlantic coast tropical moist forests of Brazil. About 90% of the former biome is still intact, much of it biologically unexplored so precluding any reliable estimate of irretrievable genetic erosion or eventual species extinction which might accompany the present deforestation of about 10 000–20 000 km²/yr (0.25–0.5% of the area). Only about 12% of the Atlantic forest area remains in original vegetation and is strongly fragmented, but no known species of its old, largely endemic fauna can be regarded as extinct. This may be due to the highly heterogeneous and naturally much-disturbed character of the Atlantic systems, preadapting the resident species to live in small populations in limited, semi-isolated habitat fragments. Various economic and social systems, leading to different models of land use, strongly influence the survival of gene pools and species during human occupation of these forest systems. Large homogeneous 'development' projects, often claimed to be sustainable, usually lead to irreversible system degradation and significant genetic erosion, whereas low-intensity extractive or mini-agricultural economies maintain or even increase the value of biological indices (endemism, diversity and protected scarce species) in these forests.

DEDICATION AND ACKNOWLEDGEMENTS

This chapter is dedicated to Dr Paulo Almeida Machado, doyen of Brazilian human ecologists, professor of public health and preventive medicine, former director of the National Amazon Research Institute (INPA) and ex-Minister of Health, on his 75th birthday, 18 July 1991.

We are especially grateful to Eduardo Brondízio of the Fundaço SOS Mata Atlântica in São Paulo, for permission to quote the results of the Atlas being developed by his group, and Roberto da Cunha and Carlos Nobre of INPE, São José dos Campos, for information on Amazonian deforestation. Work in Brazil on extractive reserves has been supported by a co-operative agreement between the UNICAMP Reitoria and the Conselho Nacional de Seringueiros, and much helped by Mauro W B Almeida and Luis Antônio Macedo. Joshua Posner has helped GGB in access to the literature on sustainable land-use methods in the tropics. In addition KSB received a fellowship from the Brazilian CNPq.

REFERENCES

Alcorn, J.B. (1989) Process as resource: the traditional agricultural ideology of the Bora and Huastec resource management and its implications for research. *Advances in Economic Botany*, 7, 63–77.

Alegretti, M.H. (1990) Extractive reserves: an alternative for reconciling development and environmental conservation in Amazonia, in *Alternatives to Deforestation: Steps Toward the Sustainable Use of the Amazon Rain Forest* (ed. A.B. Anderson), Columbia University Press, New York. pp. 252–64.

Almeida, M.W.B. (1990) Community involvement in forest management: the case of the Upper Juruá Extractive Reserve. *WWF International, Workshop on Community Involvement in Forest Management*, Denpasar, Indonesia, 15 May. 12 pp.

Bernardes, A.T., Machado, A.B.M. and Rylands, A.B. (1990) *Fauna Brasileira Ameaçada de Extinção*. Fundação Biodiversitas, Belo Horizonte. 65 pp.

Boughey, A.S. (1975) *Man and the Environment*. 2nd edn. Macmillan, New York. xi + 576 pp.

Brasil, SEPLAN/IBGE and Ministerio da Agricultura/IBDF (1988) *Mapa de vegetação do Brasil 1:5 000 000*.

Brown, K.S. Jr (1979) *Ecologia Geográfica e Evolução nas Florestas Neotropicais.* Universidade Estadual de Campinas. xxi + 265 + 120 pp.

Brown, K.S. Jr (1982) Historical and ecological factors in the biogeography of aposematic neotropical butterflies. *American Zoologist,* **44,** 453–71.

Brown, K.S. Jr (1987) Conclusions, synthesis, and alternative hypotheses, in *Biogeography and Quaternary History in Tropical America* (eds T.C. Whitmore and G.T. Prance), Oxford University Press, Oxford. pp. 175–96.

Brown, K.S. Jr (1991) Conservation of Neotropical environments: insects as indicators, in *Conservation of Insects and Their Habitats* (eds N.M. Collins and J.A. Thomas), Academic Press, London. pp. 499–504.

Brown, K.S. Jr and Cardoso A.J. (1989) *Aspectos Ecológicos da Proposta Reserva Extrativista do Tejo, Acre.* Report to Brazilian Government (also supplementary page, Conclusions and Recommendations, November). 16 pp.

Cavalcanti, D.F. (1981) Plantas em extinção no Brasil. *Boletin FBCN, Rio de Janeiro,* **16,** 115–19.

Clad, J.P. (1985) Conservation and indigenous peoples: a study in convergent interests, in *Culture and Conservation: the Human Dimension in Environmental Planning* (eds J.A. McNeely and D. Pitt), Croom Helm, London. pp. 45–60.

Colchester, M. (1989) Indian development in Amazonia: risks and strategies. *Ecologist,* **19,** 249–52.

Eccles, J C (1979) *The Human Mystery.* Springer, Berlin. xvi + 255 pp.

Endler, J.A. (1982). Pleistocene forest refuges: fact or fancy? in *Biological Diversification in the Tropics* (ed. G.T. Prance), Columbia University Press, New York. pp. 641–57.

Fearnside, P.M. (1987) Causes of deforestation in the Brazilian Amazon, in *The Geophysiology of Amazonia: Vegetation and Climate Interactions* (ed. R.E. Dickenson), Wiley, New York. pp. 37–53. (See also comment by R. Revelle and Fearnside's reply (pp. 54–61).)

Fearnside, P.M. (1989a) Extractive reserves in Brazilian Amazonia: an opportunity to maintain tropical rain forest under sustainable use. *BioScience,* **39,** 387–93.

Fearnside, P.M. (1989b) Forest management in Amazonia: the need for new criteria in evaluating development options. *Forest Ecology and Management,* **27,** 61–79.

Fearnside, P.M. Tardin, A.T. and Meira Filho, L.G. (1990) *Deforestation Rate in Brazilian Amazonia.* INPE/INPA, National Secretariat of Science and Technology. 8 pp.

Goodland, R. (1985) Tribal peoples and economic development: the human ecological dimension, in *Culture and Conservation: the Human Dimension in Environmental Planning* (eds J.A. McNeely and D. Pitt). Croom Helm. London. pp. 13–31.

Keto, A., Scott, K. and Olsen, M.F. (1990) Sustainable harvesting of tropicsl rain forests: a reassessment. Eighth session of the International Timber Council. Bali. Indonesia. 16–23 May. 5 pp.

Lovejoy, T.E., Bierregaard, R.O., Rylands, A.P., *et al.* (1986) Edge and other effects of isolation on Amazon forest fragments, in *Conservation Biology: the Science of Scarcity and Diversity* (ed. M. Soule). Sinauer. Sunderland. M.A. pp. 257–85.

McNeely, J.A., Miller, K.R., Reid, W.V., Mittermeier, R.A. and Werner, T.B. (1990) *Conserving the World's Biological Diversity*. IUCN, Gland and WRI/CI/ WWF-US/World Bank, Washington. 193 pp.

Norton, B.G. (1985) Agricultural development and environmental policy: conceptual issues. *Agriculture and Human Values*, **2**, 63–70.

Peters, C.M., Gentry, A.H., and Mendelsohn, R.O. (1989) Valuation of an Amazonian rain forest. *Nature*, **339**, 655–6.

Prescott-Allen, R. (1989) *World Conservation Strategy for the 1990s*. First Draft. 160 pp.

Redford, K.H. (1989) Monte Pascoal – indigenous rights and conservation in conflict. *Oryx*, **22**, 33–6.

Schwartz, T. (1989) The Brazilian forest people's movement. *Ecologist*, **19**, 245–7.

SOS Mata Atlântica/INPE/IBAMA (1990) *Atlas dos Remanescentes Florestais do Dominio Mata Atlântica*. Primeira Ediçã 18 maps.

Suzuki, D. (ed.) (1990) *Amazonia: the Road to the End of the Forest*. Video, Discovery/Canadian Broadcasting Corporation, 100 minutes.

Tardin, A.T., Santos, J.R. and Meira Filho, L.G. (1990) *Estado do Desflorestamento da Floresta Amazônica Brasileira em 1989*. VI. Simpósio Brasileiro de Sensoriamento Remoto, Manaus, June 24–29, 16 pp.

Thrupp, L.A. (1989) Legitimizing local knowledge: from displacement to empowerment for third world people. *Agriculture and Human Values*, **5**, 13–24.

Turner, J.R.G. (1982) How do refuges produce biological diversity? Allopatry and parapatry, extinction and gene flow in mimetic butterflies, in *Biological Diversification in the Tropics* (ed. G.T. Prance), Columbia University Press, New York. pp. 309–30.

Victor, M.A.M. (text), Cavalli, A.C., Guillaumon, J.R. and Serra Filho, R. (maps) (1975)*A Devastação Florestal*. Sociedade Brasileira de Silvicultura. 48 pp.

— 7

The influence of deforestation and selective logging operations on plant diversity in Papua New Guinea

R.J. JOHNS

INTRODUCTION

Few studies of biodiversity in tropical rain forest have been made in Papua New Guinea. From 1973 to 1977 the author conducted a detailed study in the Gogol valley in an attempt to show the influence of logging operations for chip production on the species diversity of the rain forest communities that developed after clear-felling. Fourteen 0.125 ha plots were enumerated and mapped with all species recorded (except members of the Gramineae and Zingiberaceae which were only mapped and not recorded numerically). The plots were established approximately one year after logging and were remeasured annually for four years. The study finished when the majority of sites were cleared to establish plantations of *Eucalyptus deglupta*, also for wood chips (Lamb, 1990).

In each plot (50 × 25 m) a grid was established, subdividing the plot into five 10 × 25 m subplots. For all plots a map was completed showing details of remnant logs, areas of dense regeneration and the location of seedlings and saplings. For all commercial seedlings, measurements were made of the height and diameter at breast height (Johns, 1976).

For the purposes of this biodiversity study the number of genera recorded from each plot has been averaged. These data showed the following trends over the four years (Fig. 7.1):

1. In all plots there was a rapid re-establishment of up to 60 genera of trees, shrubs, climbers and herbs within two years of logging. The sources of these were suckers and seedlings established mainly from seed existing in the seed bank, plus some contribution from seed rain. It should be noted that studies by Saulei (1985) have shown that seeds do not disperse very far from remnant forest stands. Birds which consume fruits, and act as seed disperers, seldom travel into cleared forest areas.
2. The sites supported a maximum species diversity after approximately two years, with up to 50 genera per plot.

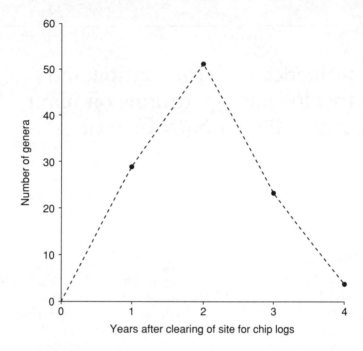

Figure 7.1. Changes in mean number of genera per plot after clear-felling of rain forest.

3. During this period aggressive secondary species, such as *Anthocephalus chinensis* and *Macaranga* spp., became established in the plots. With the dominance of this group of short-lived pioneers most other species of trees, shrubs and herbs disappeared from the site. This was presumably due to competition for light, but compounds released from the leaves or roots may have played a role.

4. After a period of three to four years the total diversity of the sites had decreased to only four or five woody species per 0.125 hectare plot (Fig. 7.1).

Evidence from areas which were burnt in forest fires shows that the re-establishment of a diverse forest will be very slow (Johns, 1983). In the South Naru area which was burnt in fires in 1941, the river terraces (reportedly burnt) were very species-poor in comparison to the hill forest (Johns, 1983). Taylor (1954) estimated a time lapse of some 600 years for re-establishment of a species-diverse, 'self-perpetuating', tropical rain forest following destruction by volcanic activity. The forests of the Gogol valley are likely to have been severely effected by the eruptions of Long Island some 300 years ago (Johns, 1983).

SELECTIVE LOGGING

The Gogol area is one of three dryland forest wood chipping operations in the humid tropics (Whitmore, 1990). Most forest operations in Papua New Guinea and elsewhere in the tropics use selective logging techniques.

The effects of selective logging on species richness are difficult to assess. Tropical rain forest in Papua New Guinea has a very low density of tree species which have timber accepted on the international market. Consequently the volumes cut for logs in most operations range from 20 to 30 m³/ha, and seldom exceed 45 m³/ha. Thus in a typical selective logging operation only five to six trees are removed per ha. This volume of extracted timber is typical of many dipterocarp forests in western Malesia, although some forests in Borneo yield up to 120 m³/ha.

Unfortunately, logging has two major impacts on the forest.

1. The physical effects of the removal of logs includes damage during felling operations, damage during skidding, clearing for log dumps and the destruction of drainage systems during track construction. In addition, microclimatic changes can occur when extensive areas of the canopy are disturbed. This will probably adversely affect epiphytes, particularly in high altitude logging operations where the greatest diversity of epiphytic plants occurs.
2. Logging operations which provide access into previously remote areas (where the richest forest resources occur) are commonly followed by land clearance for farming. Most post-logging development occurs for agriculture despite the very poor export values of, or prospects for, most tropical crops such as cocoa, coffee, rubber, oil palm and copra. Agriculture, particularly the establishment of plantation crops, is still promoted by the Government of Papua New Guinea in spite of the economic, social or biological problems associated with deforestation.

Several areas in central New Britain were recently studied by a group from the PNG University of Technology in Lae (Arentz *et al.*, 1989) to assess the impact of the selective logging operations. A list of species previously collected from New Britain was prepared from the available literature (Johns in Arentz *et al.*, 1989) and from previous collections held in the National Herbarium at Lae. It is not possible to assess adequately the effects of the selective logging operations on species diversity. After logging the total destruction of large forest areas had occurred as a consequence of the establishment of oil palm plantations. Areas were examined where the forests had not been disturbed after logging. These forests were quite diverse in species composition and included many species of filmy ferns (Hymenophyllaceae), a family that perhaps indicates that the microclimatic effects of the operation were not as extreme as one might expect. Several previously unrecorded species were collected from the logged areas, this probably reflects low prelogging collecting intensities rather than new species becoming established as a consequence of logging. The group of species invading after logging included both short-lived and long-lived pioneers. Plot data collected in unlogged forest showed that the important commercial species were poorly represented in the subcanopy of the forest (Arentz *et al.*, 1989), perhaps indicating that this forest is in a seral state. In regrowth forests seedlings of some commercial species were found to be regenerating in canopy openings caused by logging.

Johns, (1977) has previously noted the seral nature of the *Castanopsis* forests of the lower montane zone in Papua New Guinea. It appears probable that excluding logging or providing strict protection through establishment of national parks or nature reserves would result in the development of the Gogol Valley forests into mixed forests of quite different species composition from those found in the area today (Johns, 1977).

CONCLUSIONS

Comparatively little is known of the effects of logging on plant or animal species richness in Papua New Guinea. A single cycle of selective logging does not appear to diminish plant diversity and does favour regeneration of commercially valuable timber species. Subsequent clearance of the forest for agriculture is a serious problem. Selective logging could be an acceptable technique for the sustained yield management of lowland tropical rain forest in Papuasia.

SUMMARY

Detailed studies were undertaken in the Gogol valley, Papua New Guinea, to assess rain forest regeneration following clear-felling for wood-chip production and selective logging for export-grade lumber. Clear-felling resulted in almost complete loss of the natural speices diversity of the forest. In comparison, a single cycle of selective logging, at least when practised in Papua New Guinea, has only a minor effect on biodiversity. The major problems of habitat destruction are associated with the complete destruction of rain forest communities for agricultural development, in areas where logging infrastructure provides access.

REFERENCES

Arentz, F., Johns, R.J., Lamothe, L., Matcham, E.J., Simaga, J. and Taurereko, R. (1989) The forests of New Britain: Central New Guinea. *PNG University of Technology* Lae, Papua New Guinea.

Johns, R.J. (1976) Natural vegetation following chip logging in the tropical rain forest: a case study from the Gogol river valley. Paper presented at the Third Meeting of the Papua New Guinea Botanical Society, Lae.

Johns, R.J. (1977) Habitat conservation in Papua New Guinea. Unpublished paper. PNG Botanical Society.

Johns R.J. (1983) The instability of the tropical ecosystem in New Guinea. *Blumea*, **31**, 341–71.

Johns, R.J. (1990) The illusionary concept of the climax, in *The Plant Diversity of Malesia* (eds P. Baas, R. Geesink and K. Kalkman), Leiden.

Lamb, D. (1990) Exploiting the tropical rain forest. An account of pulpwood logging in Papua New Guinea. *UNESCO*, Paris.

Saulei, S. (1984) Natural regeneration following clear-fell logging operations in the Gogol Valley, Papua New Guinea. *Ambio*, **13**, 351–4.

Saulei, S. (1985) The recovery of lowland rain forest after clear-fell logging operations in the Gogol Valley, Papua New Guinea. PhD thesis. University of Aberdeen.

Taylor, P.W. (1954) Plant succession on recent volcanoes in Papua. *Journal of Ecology*, **45**, 233–43.

Whitmore, T.C. (1990) *Introduction to Tropical Rain Forests*. Clarendon Press, Oxford.

Whyte, I.N. (1977) Results of a regeneration survey in the Gogol. Paper presented at the PNG Botanical Society, Bulolo.

Index

Those entries in *italics* represent figures and those in **bold** are tables.